Practical Problems in Mathematics
FOR Printers

Practical Problems in Mathematics

FOR Printers

JAMES P. DE LUCA

DELMAR PUBLISHERS
COPYRIGHT © 1976
BY LITTON EDUCATIONAL PUBLISHING, INC.

All rights reserved. No part of this work covered by the copyright hereon may be reproduced or used in any form or by any means — graphic, electronic, or mechanical, including photocopying, recording, taping, or information storage and retrieval systems — without written permission of the publisher.   10 9 8 7 6 5 4 3 2 1

LIBRARY OF CONGRESS CATALOG CARD NUMBER: 76-3942

Printed in the United States of America
Published Simultaneously in Canada by
Delmar Publishers, A Division of
Van Nostrand Reinhold, Ltd.

DELMAR PUBLISHERS • ALBANY, NEW YORK 12205
A DIVISION OF LITTON EDUCATIONAL PUBLISHING, INC.

PREFACE

PRACTICAL PROBLEMS IN MATHEMATICS FOR PRINTERS is one in a series of workbooks designed to be used in conjunction with BASIC MATHEMATICS SIMPLIFIED by C. Thomas Olivo and Thomas P. Olivo (Delmar Publishers) to provide practical experiences in solving occupationally related problems using basic mathematics principles.

The previous edition of PRACTICAL PROBLEMS IN MATHEMATICS FOR PRINTERS emphasized linecasting machines such as the Linotype. The present edition reflects the universal acceptance of the much more rapid composition processes in which no hot metal is used. These modern processes are known by several descriptive phrases such as cold type, phototypesetting, photocomposition, flat type, or nonmetallic composition.

The 1976 edition of this workbook was broadened in scope to provide the student with practical examples of the interrelationships between all of the working departments of a printing establishment. The terminology used throughout the workbook emphasizes the relationships of the specialty areas, such as copy, typography, camera, stripping, and presswork, to the craft of printing.

The majority of the problems concerns nonmetallic composition and the offset method of lithographic printing. All of the problem material has been classroom tested with graphic arts and printing students. This workbook will provide students, whether in secondary vocational, post secondary, apprentice training, or adult education programs, with the practice necessary to obtain a basic competence in solving printing related problems. In view of the forthcoming conversion by the United States to the metric system, several units provide practice in using metric units.

Two summary Achievement Reviews are provided after the basic problem units. Each of these reviews tests the principles presented and practiced throughout the entire workbook. An Instructor's Guide is provided and includes the answers to all of the problems as well as the answers to the Achievement Reviews.

James P. DeLuca presently is the Chairman of the Department of Graphic Arts and Advertising Technology and the Lithographic Offset Technology Program of the New York City Community College of the City University of New York.

CONTENTS

SECTION 1 WHOLE NUMBERS

Unit

1	Addition of Whole Numbers	1
2	Subtraction of Whole Numbers	4
3	Multiplication of Whole Numbers	6
4	Division of Whole Numbers	7

SECTION 2 FRACTIONS

5	Addition of Fractions	8
6	Scale Reading	11
7	Subtraction of Fractions	13
8	Multiplication of Fractions	15
9	Division of Fractions	17

SECTION 3 DECIMALS

10	Addition of Decimals	19
11	Subtraction of Decimals	22
12	Multiplication of Decimals	25
13	Division of Decimals	26
14	Reading an Outside Micrometer	28

SECTION 4 PERCENT

15	Fractional Equivalents	29
16	Simple Percent	31
17	Discounts	33
18	Profit and Loss, Commissions	35
19	Interest and Taxes	37

SECTION 5 MEASUREMENT

20	Linear Measure	39
21	Metric System	41
22	Metric Equivalents	43

Contents

23	Angular Measurement	44
24	Units of Area and Volume Measure	46
25	Time and Money Calculations	48

SECTION 6 GRAPHS

26	Practical Applications of Graphs and Charts	49

SECTION 7 PAPER STOCK

27	Packaging Paper Stock	53
28	Basic Size, Thickness, and Weight of Stock	56
29	Equivalent Weights of Stock	59
30	Determining and Cutting Paper Stock	63
31	Determining the Number of Sheets Required	66
32	Finding the Most Economical Cut of Stock	71
33	Allowance for Paper Spoilage	73
34	Charging for Cutting and Handling Stock	76
35	Determining Weight of Paper Stock	78
36	Determining Cost of Paper Stock	80

Achievement Review A	86
Achievement Review B	89
Answers to Odd-Numbered Review Problems	93
Acknowledgement Page	98

The author and editorial staff at Delmar Publishers are interested in continually improving the quality of this instructional material. The reader is invited to submit constructive criticism and questions. Responses will be reviewed jointly by the author and source editor. Send comments to:

Director-in-Chief
Box 5087
Albany, New York 12205

SECTION 1 — WHOLE NUMBERS

UNIT 1 ADDITION OF WHOLE NUMBERS

BASIC PRINCIPLES OF ADDITION

- Study Unit 1 in *Basic Mathematics Simplified* for the principles of addition as applied to whole numbers.
- Apply the principles of addition to the printing and graphic communications industry by solving the review problems.

REVIEW PROBLEMS

1. A printer completes the following jobs in one day: 6,750 letterheads, 7,250 business cards, 17,250 circulars and 12,675 labels. Determine the total number of pieces of work the printer completes during the day.   _____

2. Determine the total of the following payroll: _____

2 offset pressmen	$500.00
1 cameraman	$250.00
1 stripper	$275.00
1 platemaker	$225.00

3. The following lengths of lithographic film are cut from a roll of film: 17 inches, 19 inches, 24 inches, 18 inches, 12 inches, 21 inches and 15 inches. How many inches of film are cut from the roll?   _____

4. At the XYZ Lithographic Printing Company there are 17 office workers, 4 pressmen, 6 cameramen, 3 strippers, 2 platemakers, 3 commercial artists; 1 plant manager and 2 foremen. Determine the number of people working at this printing company.   _____

5. What is the combined weight of the four machines shown in Fig. 1-1?   _____

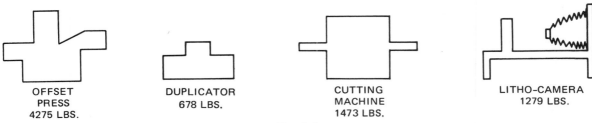

OFFSET PRESS 4275 LBS.   DUPLICATOR 678 LBS.   CUTTING MACHINE 1473 LBS.   LITHO-CAMERA 1279 LBS.

Fig. 1-1

6. Fig. 1-2, page 2, shows the ink used on each of six offset presses during three work shifts. Determine the total amount of ink used during the three shifts, in pounds (lbs.).   _____

Unit 1 — Addition of Whole Numbers

WORK SHIFT	OFFSET PRESSES-INK IN POUNDS						TOTAL FOR SHIFT
	1	2	3	4	5	6	
One	22	15	12	22	11	17	?
Two	17	12	13	27	12	28	?
Three	15	18	19	28	9	14	?
	TOTAL FOR THREE SHIFTS						?

Fig. 1-2

7. In Fig. 1-3, the time card shows the hours spent producing a printing job. Determine the total amount of time required for this job.

TIME CARD	
AREA	HOURS
Camera	14
Stripping	19
Platemaking	3
Pressroom	35
Cutting	11
TOTAL HOURS	?

Fig. 1-3

8. A monthly magazine, during a six-month period, distributes the following number of copies: 16,200; 21,500; 24,500; 30,000; 32,500; 35,000. Determine the total number of copies the magazine distributes during this six-month period.

9. A daily newspaper delivers the following number of papers to five rural villages: village "A", 1,425 papers; village "B", 925 papers; village "C", 725 papers; village "D", 870 papers; village "E", 1,100 papers. What is the total daily delivery to these five villages?

10. A publisher plans to deliver books in four separate shipments as follows: shipment "A", 6,750 books; shipment "B", 9,750 books; shipment "C", 11,250 books; shipment "D", 14,300 books. What is the total number of books the publisher plans to deliver?

11. The following items are in an offset stripping tool box: 6 boxes of razor blades, 4 sets of pins and fitters, 3 straight edges, 13 rolls of tape and 6 rulers. Find the total number of items in the box.

12. Five pieces of paper are cut from a roll of blueprint paper. The five pieces are: 13 inches, 19 inches, 11 inches, 24 inches, and 17 inches long. Determine the total length of material cut from the roll.

13. a. In Fig. 1-4, determine distance A, in inches. _____
    b. In Fig. 1-4, determine distance B, in inches. _____
    c. In Fig. 1-4, determine distance C, in inches. _____

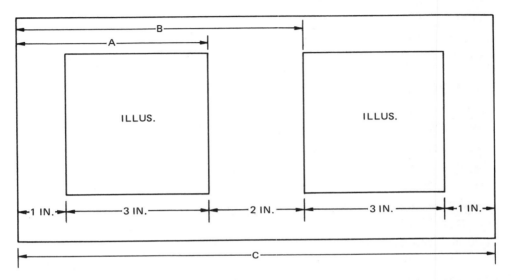

Fig. 1-4

14. Fig. 1-5 shows the cost of printing each part of a brochure. Determine the total cost of the job. _____

Artwork	$ 125.00
Camera and Stripping	375.00
Platemaking	210.00
Presswork	1,240.00
Cutting/Shipping	135.00
TOTAL COST	?

    Fig. 1-5

15. A typographer sets 1,575 lines of type on Monday; 2,200 lines of type Tuesday; and 1,850 lines of type on Wednesday. Determine the total number of lines the typographer sets during the three days. _____

UNIT 2 SUBTRACTION OF WHOLE NUMBERS

BASIC PRINCIPLES OF SUBTRACTION

- Study Unit 2 in *Basic Mathematics Simplified* for the principles of subtraction as applied to whole numbers.
- Apply the principle of subtraction to the printing and graphic communications industry by solving the review problems.

REVIEW PROBLEMS

1. If seventeen gallons of presswash are removed from a drum containing fifty-five gallons of presswash, how many gallons are left in the drum?

2. a. In box 1 there are 275 pieces of 8-point type, 325 pieces of 10-point type, 427 pieces of 12-point type and 129 pieces of 18-point type. Determine the number of pieces of type in box 1.

   b. In box 2 there are 112 pieces of 8-point type, 629 pieces of 10-point type, 318 pieces of 12-point type and 121 pieces of 18-point type. Determine the number of pieces of type in box 2.

   c. There are more pieces of type in box 2 than box 1. Determine how many more pieces of type are in box 2.

3. During a month, camera one uses $335.00 of electricity while camera two uses $227.00 of electricity. How much more electricity does camera one use than camera two?

4. If a publisher purchases a carload of paper stock for $39,600.00 and sells it to another publisher for $42,350.00, determine the amount of profit the first publisher makes.

5. a. Determine length A in Fig. 2-1, in inches.
   b. Determine length B in Fig. 2-1, in inches.

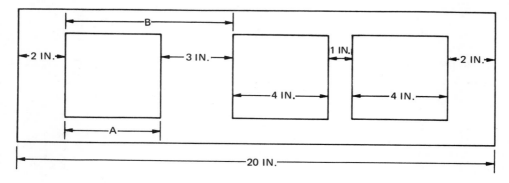

Fig. 2-1

6. Three pieces of photographic film measuring 23 inches, 19 inches and 37 inches long are cut from a roll measuring 125 inches long. How much is left on the original roll in inches?

Section 1 — Whole Numbers

7. A typographer works thirty-five hours during a week. Four hours are spent doing office corrections and three hours are spent in checking out the equipment. These hours are not chargeable to a job. If the remaining hours are chargeable to a job, how many hours are chargeable? _____

8. A printer charges an advertising agency $3,250.00 for a printing job and in turn the advertising agency charges their client $4,125.00 for the job. What is the amount of profit the advertising agency makes? _____

9. If pressman "A" runs 22,500 impressions, and pressman "B" runs 27,300 impressions, how many more impressions does pressman "B" run than pressman "A"? _____

10. The estimated value of a small offset plant was $11,250.00. Since the estimate, the value has decreased by $1,200.00. Determine the present value of the plant. _____

UNIT 3 MULTIPLICATION OF WHOLE NUMBERS

BASIC PRINCIPLES OF MULTIPLICATION

- Study Unit 3 in *Basic Mathematics Simplified* for the principles of multiplication as applied to whole numbers.
- Apply the principles of multiplication to the printing and graphic communications industry by solving the review problems.

REVIEW PROBLEMS

1. A litho-offset printing press can produce 3,500 impressions per hour. How many impressions can be produced if the press runs for six hours? _____

2. A printer buys 39,000 pounds of offset paper stock at $.59 a pound. How much does she pay for the paper stock? _____

3. There are 2,100 pounds of paper on one skid. Determine the total weight in pounds of the paper on five skids. _____

4. A book of 128 pages contains 22 lines on each page. How many lines are there in the book? _____

5. If four letterheads can be cut from a single sheet, how many letterheads can be cut from 8,750 sheets? _____

6. One pamphlet weighs three ounces. A printer prints 22,000 pamphlets. Determine the total weight of the pamphlets, in ounces. _____

7. A publisher purchases three carloads of paper stock, each carload weighing 42,000 pounds. What is the total weight of the three carloads? _____

8. There are 500 sheets of paper in a ream of bond paper. Determine the number of sheets in 26 reams of bond paper. _____

9. One ream of book paper weighs 16 pounds. Determine the weight of 22,500 reams of book paper. _____

10. A typesetter can set an average of 320 lines of type per hour. Determine the number of lines of type the typesetter can set in seven hours. _____

UNIT 4 DIVISION OF WHOLE NUMBERS

BASIC PRINCIPLES OF DIVISION

- Study Unit 4 in *Basic Mathematics Simplified* for the principles of division as applied to whole numbers.
- Apply the principle of division to the printing and graphic communications industry by solving the review problems.

REVIEW PROBLEMS

1. A ream of paper contains 500 sheets of paper. In a stockroom there are 12,500 sheets of paper. Determine the number of reams in the stockroom. _____

2. How many direct image offset printing plates, each 15 inches long, can be cut from a roll which is 255 inches long? _____

3. A box of 50 sheets of litho film costs $32.00. What is the cost of each sheet of film? _____

4. A publisher orders 62,750 pamphlets which are delivered in five equal shipments. Determine the number of pamphlets in each shipment. _____

5. A book to be printed has a total of 2,278 lines of type. How many pages will there be in the book if there are 32 lines of type on each page? _____

6. If a pressman makes $52.00 each day, how many days does it take to earn $624.00? _____

7. A printing job is done by fifteen people. The job takes 450 hours to complete. If each person works an equal amount on the job, determine the number of hours each person works. _____

8. An order for 25,000 direct mail pieces is placed by an advertising agency. The order has a total weight of 100,000 ounces. What is the weight of each piece? _____

9. A shipping box can hold 325 pieces of printed material. A printer wishes to ship 44,850 pieces of printed material. Determine the number of boxes needed to ship the 44,850 pieces. _____

10. Two photo-typesetters work a total of 350 hours on a job. If they work 7 hours a day, 5 days a week, how many weeks does the job take? _____

Section 2 — FRACTIONS

UNIT 5 ADDITION OF FRACTIONS

BASIC PRINCIPLES OF ADDITION

- Study Unit 7 in *Basic Mathematics Simplified* for the principles of addition as applied to fractions.
- Apply the principles of addition to the printing and graphic communications industry by solving the review problems.

REVIEW PROBLEMS

1. An offset printing press is run 3 3/4 hours on Monday, 6 1/4 hours on Tuesday, 7 hours on Wednesday, 4 1/2 hours on Thursday and 5 1/4 hours on Friday. Determine the total number of hours the press is run during the five days.

2. A pressman uses the following amounts of ink on a job: 10 1/2 pounds of blue, 16 pounds of red, 15 1/4 pounds of yellow, and 9 3/4 pounds of black. Determine the total number of pounds of ink the pressman uses for the job.

3. The inventory of a small print shop shows the following paper stock on hand: 12 1/2 reams of bond paper, 27 1/2 reams of mimeograph paper and 29 1/2 reams of offset duplicator paper. Determine the total number of reams of paper in the inventory.

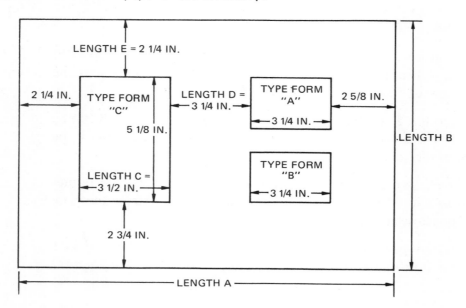

Fig. 5-1

Section 2 — Fractions

4. a. Determine length A in Fig. 5-1, in inches. _____

    b. Determine length B in Fig. 5-1, in inches. _____

    c. If length C is 3 1/8 in. instead of 3 1/2 in., determine length A in Fig. 5-1, in inches. _____

    d. If length C is 2 7/8 in. instead of 3 1/2 in., determine length A in Fig. 5-1, in inches. _____

    e. If length D is 2 5/8 in. instead of 3 1/4 in., determine length A in Fig. 5-1, in inches. _____

    f. If length E is 1 7/8 in. instead of 2 1/4 in., determine length B in Fig. 5-1, in inches. _____

5. A cameraman uses the following amounts of developer in one day: 3 1/2 gallons of type "A", 2 1/4 gallons of type "B", 6 3/4 gallons of type "C". Determine the total amount of developer the cameraman uses during the day. _____

6. An advertising agency orders a full page advertisement for a magazine shown in Fig. 5-2. Using the following dimensions, determine the total height of the page in inches (in.): 1 3/4 inches for the top margin, 5 7/8 inches for an illustration and 1 3/8 inches for the bottom margin. _____

7. A printer receives copy from a customer which calls for four ruled columns. If these columns are to be 3/4 inches, 1 1/8 inches, 2 5/8 inches and 9/16 inches wide, determine the total width of the copy in inches. _____

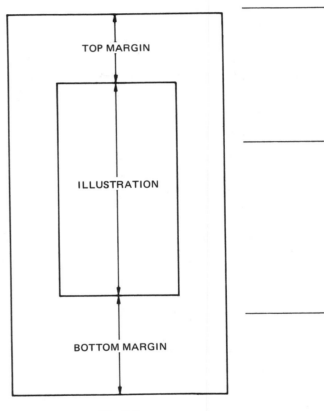

Fig. 5-2

8. A stripper works 3 1/2 hours on one job, 3 1/4 hours on another job and 3/4 hour on a third job. Determine the total amount of time the stripper works on the three jobs. _____

9. Fourteen and one-half hours are required for camera, stripping and platemaking on a job, nine and one-quarter hours for presswork and one and one-half hours for cutting. Determine the total number of hours needed to complete this job. _____

Unit 5 — Addition of Fractions

10. The time sheet for printing a job, shown in Fig. 5-3, shows the amount of time needed for each part of the job.

    a. Determine the total amount of time needed for the job. _____

    b. Determine the total amount of time needed for the job if 6 3/4 additional hours are also needed for the artwork for the job. _____

TIME SHEET	
Camera	8 1/4 hours
Stripping & Platemaking	14 3/4 hours
Presswork	21 1/2 hours
Cutting & Shipping	9 1/4 hours
Total Time	?

    Fig. 5-3

11. An apprentice's weekly classroom schedule is made up as follows: 4 1/2 hours in a camera class, 2 1/2 hours in a class on stripping and 1 1/4 hours in a platemaking class. Determine the number of hours the apprentice spends in the three classes. _____

12. Three artists work on the design of a new package for their advertising agency. The first artist works 6 1/4 hours; the second works 7 1/2 hours; and the third works 5 3/4 hours. Determine the total number of hours the three artists work on this job. _____

13. During one week, a cameraman uses 22 1/2 gallons of developer, 14 1/4 gallons of fixer and 2 1/2 gallons of film cleaner. Determine the total amount of photographic chemicals this cameraman uses during the week. _____

14. The inventory of a stripping department consists of the following number of packages of stripping material: 3 1/2 reams, small size; 23 3/4 reams, medium size; and 33 1/4 reams, large size. Determine the total number of reams of stripping material in the inventory. _____

UNIT 6 SCALE READING

BASIC PRINCIPLES OF SCALE READING

- Study Unit 6 in *Basic Mathematics Simplified* for the principles of scale reading as applied to fractions.

- Apply the principles of scale reading to the printing and graphic communication industry by solving the review problems.

REVIEW PROBLEMS

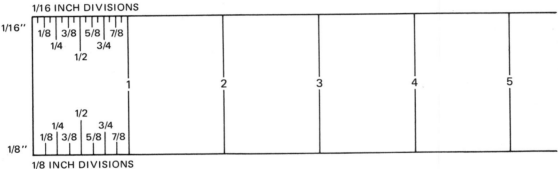

Fig. 6-1 Six-inch Ruler

1. On the upper part of the ruler shown in Fig. 6-1 each division equals 1/16 inch. This means there are 16 divisions in each inch. How many 1/16-inch divisions are there in 3 inches? _____

2. On the lower part of the ruler shown in Fig. 6-1 each division equals 1/8 inch. This means there are 8 divisions in each inch. How many 1/8-inch divisions are there in 3 inches? _____

3. How many 1/16-inch divisions are there in 1/4 inch? _____

4. How many 1/16-inch divisions are there in 3/8 inch? _____

5. How many 1/16-inch divisions are there in 7/8 inch? _____

6. How many 1/8-inch divisions are there in 3/4 inch? _____

7. How many 1/8-inch divisions are there in 1/4 inch? _____

8. How many 1/8-inch divisions are there in 1/2 inch? _____

11

Unit 6 — Scale Reading

9. Give the scale readings indicated on the ruler shown in Fig. 6-2, in inches.

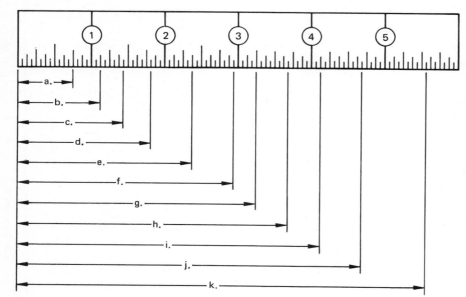

Fig. 6-2

a. _____
b. _____
c. _____
d. _____
e. _____
f. _____
g. _____
h. _____
i. _____
j. _____
k. _____

10. Using a ruler, measure the following lines to the nearest 1/16 inch:

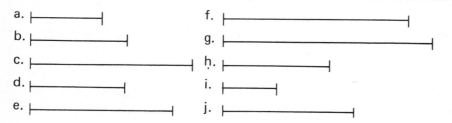

a. _____
b. _____
c. _____
d. _____
e. _____
f. _____
g. _____
h. _____
i. _____
j. _____

Fig. 6-3

11. With a ruler, determine the total lengths of line A, line B, and line C in Fig. 6-3, in inches.   _____

12. With a ruler, determine the total lengths of line D, line E, and line K in Fig. 6-3, in inches.   _____

13. With a ruler, determine the total lengths of line F, line I, and line J in Fig. 6-3, in inches.   _____

UNIT 7 SUBTRACTION OF FRACTIONS

BASIC PRINCIPLES OF SUBTRACTION

- Study Unit 8 in *Basic Mathematics Simplified* for the principles of subtraction as applied to fractions.
- Apply the principles of subtraction to the printing and graphic communications industry by solving the review problems.

REVIEW PROBLEMS

1. From a roll of direct image platemaking material 250 inches long, three pieces are cut. The first piece is 17 1/2 inches long, the second piece is 23 1/4 inches long and the third piece is 19 3/4 inches long. Determine the amount of material remaining on the roll, in inches.

2. A skid of paper stock contains 52 1/2 reams of paper. A pressman uses 29 1/4 reams to print a job. Determine the amount of paper remaining on the skid, in reams.

3. A small offset duplicating plant has 3/4 ream of offset paper in stock. For a small job, 1/3 ream of offset paper is used. Determine the amount of paper remaining in stock.

4. A brochure requires 39 3/4 hours to complete while a second brochure takes 53 1/4 hours to complete. How much more time is needed to complete the second brochure?

5. A printer estimates that a job will require 79 1/2 hours to print. The actual time cards show the job requires 85 1/4 hours to print the job. Determine the number of hours the job requires over the estimate given by the printer.

6. Fig. 7-1 shows a piece of paper with an illustration mounted on it.
    a. Determine the width of the right-hand margin on the paper, length A, in inches.
    b. If the width of the paper, length B, is 9 1/4 inches instead of 8 1/2 inches, determine the width of the right-hand margin on the paper, length A, in inches.
    c. If the width of the illustration is 4 3/4 inches instead of 5 1/2 inches, and the width of the paper, length B, is 8 1/2 inches, determine length A, in inches.
    d. Determine the height of the illustration, length C, in inches.
    e. If the height of the paper, length D, is 11 3/4 inches instead of 12 1/4 inches, determine length C, in inches.

7. A can contains 10 1/2 pounds of ink; for a job, 3 3/4 pounds of ink are used. Determine the weight of the ink remaining in the can.

Unit 7 — Subtraction of Fractions

8. A sheet of paper contains 787 1/2 square inches of space. An illustration covers 539 1/4 square inches of space. Determine the amount of paper not covered by the illustration, in square inches.

9. The following amounts of presswash are taken from a 55-gallon drum: 5 1/4 gallons, 3 3/4 gallons, 1 1/2 gallons, and 6 1/4 gallons. Determine the amount of presswash remaining in the drum, in gallons.

10. The time required to complete a job is 7 1/2 hours. A person works on the job during three different days as follows: 3/4 hour during the first day, 1 1/4 hours during the second day, and 2 1/4 hours during the third day. How many hours are needed to complete the job?

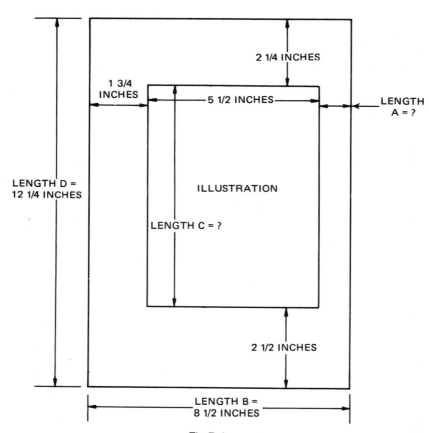

Fig. 7-1

UNIT 8 MULTIPLICATION OF FRACTIONS

BASIC PRINCIPLES OF MULTIPLICATION

- Study Unit 9 in *Basic Mathematics Simplified* Third Edition for the principles of multiplication as applied to fractions.
- Study Unit 21, Section A and Section C, in *Basic Mathematics Simplified* Third Edition for the principles of area measurement; or

  Study Unit 20, Section A and Section C, in *Basic Mathematics Simplified* **Fourth Edition** for the principles of area measurement.
- Apply the principles of multiplication to the printing and graphic communications industry by solving the review problems.

REVIEW PROBLEMS

1. A printing plant purchases 12 1/2 pounds of blue process ink at a cost of $.82 per pound. What is the total cost of the ink?
2. What is the weight of a 55-gallon drum of varnish if each gallon weighs 5 1/4 pounds?
3. What is the weight of 14 reams of paper if each ream weighs 21 1/4 pounds?
4. Determine the width of a newspaper page if the page contains 5 columns and each column is 2 1/8 inches wide.
5. An offset pressman prints 4,250 sheets of paper an hour for 7 1/2 hours. Determine the total number of sheets the pressman prints.
6. A typesetter sets 6 columns of type per hour. Determine the number of columns the typesetter sets in 3 1/4 hours.
7. A line of 8-point type has an average of 10 1/2 words per line. How many words are there in 118 lines of type?
8. An order calls for the finished book to be 8 1/2 inches x 10 3/4 inches with a 5/8-inch spine as shown in Fig. 8-1.
   a. If 3/4 inch is allowed on all sides for binding of the cover, determine the total area of one cover in square inches.
   b. There are 5,500 covers needed for the order. Determine the total area of the 5,500 covers, in sq. in.

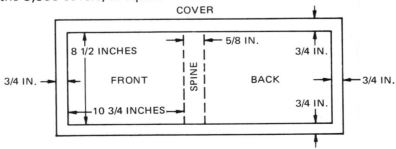

Fig. 8-1

Unit 8 — Multiplication of Fractions

9. Six pieces of paper, each 4 3/4 inches long, are needed for a printing job. Determine the total length of the six pieces, in inches. _____

10. Determine the area of Illustration B in Fig. 8-2, in square inches. _____

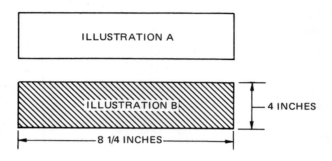

Fig. 8-2

11. Six lengths of photo-typesetting film each 7 1/4 inches long are cut from a roll of film. How many inches of film are cut from the roll? _____

12. A folding machine can fold 4,000 sheets of paper in an hour. How many sheets will be folded if the machine is run for 5 1/4 hours? _____

UNIT 9 DIVISION OF FRACTIONS

BASIC PRINCIPLES OF DIVISION

- Study Unit 10 in *Basic Mathematics Simplified* for the principles of division as applied to fractions.
- Apply the principles of division to the printing and graphic communications industry by solving the review problems.

REVIEW PROBLEMS

1. An offset pressman prints 38,000 sheets in 9 1/2 hours. What is the average number of sheets the pressman prints per hour?

2. A folding machine is run for 12 1/2 hours and folds 50,000 sheets of paper. What is the average output per hour?

3. During a work shift of 7 1/2 hours, 127.5 feet of gold leaf material is used. What is the average amount of gold leaf used per hour?

4. How many pieces of lithographic film, each 6 1/2 inches long, can be cut from a roll 100 inches long?

5. a. Determine length X in Fig. 9-1, in inches.
   b. Determine height Y in Fig. 9-1, in inches.

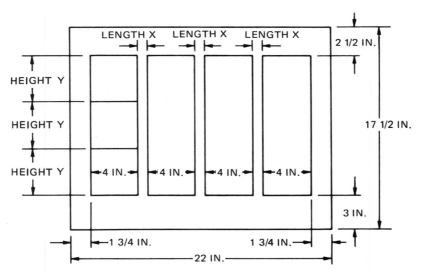

Fig. 9-1

6. A page of type is 6 3/4 inches wide and is divided into three equal columns. How many inches wide is each column?

7. a. The length of the bristol board is 19 1/2 inches. How many pieces, each 4 1/2 inches long, can be cut from the length of bristol board?
   b. Determine the amount of bristol board remaining after the maximum number of pieces are cut from the board.

Unit 9 — Division of Fractions

8. A typesetter can set an average of 56 lines of type each hour. If there are 1,078 lines of type to be set for a job, how many hours will be needed to set the job?

9. In Fig. 9-2, determine length A, the width of each column, in inches. The widths of the three columns are all equal.

10. If a camera department uses 45 gallons of fixer in 3 days, what is the average amount of fixer the department uses each day?

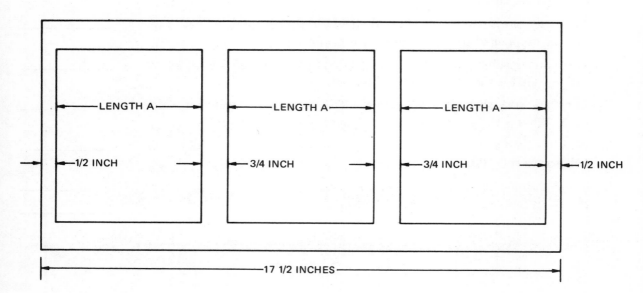

Fig. 9-2

SECTION 3 — DECIMALS

UNIT 10 ADDITION OF DECIMALS

BASIC PRINCIPLES OF DECIMALS

- Study Unit 13 in *Basic Mathematics Simplified* for the principles of addition as applied to decimals.
- Apply the principles of addition to the printing and graphic communications industry by solving the review problems.

REVIEW PROBLEMS

1. During a week, a commercial artist works the following amount of time on a project:

   Monday.......................... 6.5 hours
   Tuesday......................... 4.2 hours
   Thursday........................ 5.3 hours

   Determine the total number of hours the artist works on the project.

2. A shipment of photographic chemicals is received. The weight of each chemical is listed:

   Litho developer.......................... 37.5 pounds
   Replenisher for Automatic Processors .... 72.3 pounds
   Hypo (fixer) .... 41.6 pounds

   Determine the total weight of the shipment.

3. Fig. 10-1 shows a book. The front and back covers of the book are each 0.0746 inch thick. The stock measures 0.9171 inch. Determine the thickness, T, of the entire book.

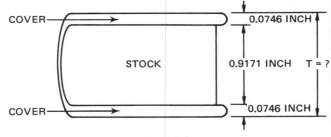

Fig. 10-1

4. The working time of a pressman during one week is shown:

Monday	4.7 hours	Thursday	3.8 hours
Tuesday	5.6 hours	Friday	4.9 hours
Wednesday	7.1 hours		

   Determine the total working time of the pressman.

Unit 10 — Addition of Decimals

5. The advertisment shown in Fig. 10-2 has a margin at the top of 0.75 inch. There are 3.7 inches of illustrations above 6.1 inches of reading matter. Determine the height, H, of the advertisement. _____

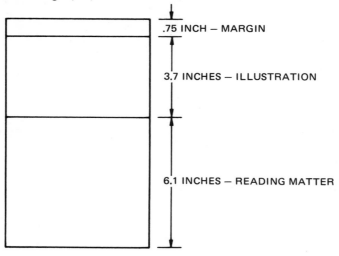

Fig. 10-2

6. The time a worker spends preparing a flat for the platemaker is listed: _____
   Retouching . . . . 1.7 hours
   Stripping . . . . 3.5 hours
   Contacts . . . . 0.6 hours

   Determine the total time the worker spends preparing the flat.

7. a. Determine the overall length, length C, of the form roller shown in Fig. 10-3. _____

   b. If length A is 3.175 inches and length B is 29.750 inches, determine length C, in Fig. 10-3. _____

   c. If length A is 2.760 inches and length B is 31.725 inches, determine length C in Fig. 10-3. _____

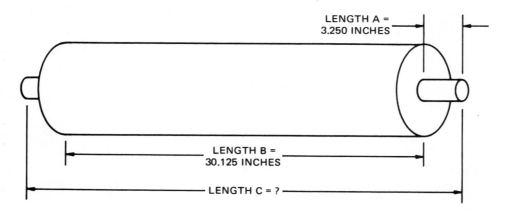

Fig. 10-3

20

Section 3 — Decimals

8. Determine the total amount of the following bill:

Type and artwork.	$ 17.75
Camera	22.50
Stripping.	37.40
Platemaking.	49.25
Presswork	325.75
Finishing.	52.50
Tax	<u>37.75</u>
TOTAL	$

9. A customer buys one can of black ink for $1.75, one can of blue ink for $2.25, and one can of red ink for $2.75. Determine the total cost of the three cans of ink.

10. The following time is spent in the platemaking department:

Monday	7.6 hours
Tuesday.	3.7 hours
Wednesday	6.5 hours

    Determine the total amount of time spent working in the platemaking department.

11. The weekly expense of a messenger are listed:

Railroad	$13.75
Subway.	1.40
Taxi	7.25
Bus	2.80

    Determine the total expense of the messenger.

UNIT 11 SUBTRACTION OF DECIMALS

BASIC PRINCIPLES OF SUBTRACTION

- Study Unit 14 in *Basic Mathematics Simplified* for the principles of subtraction as applied to decimals.
- Apply the principles of subtraction of decimals to the printing and graphic communications industry by solving the review problems.

REVIEW PROBLEMS

1. The collar shown in Fig. 11-1 has an outside diameter, O.D., of 1.7246 inches and a wall thickness, W.T., of 0.327 inch. Determine the inside diameter, I.D., of the collar.

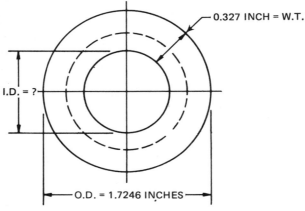

Fig. 11-1

2. Determine the inside diameter of the bushing shown in Fig. 11-2 if the outside diameter, O.D., is 2.765 inches and the wall thickness, W.T., is 0.794 inch.

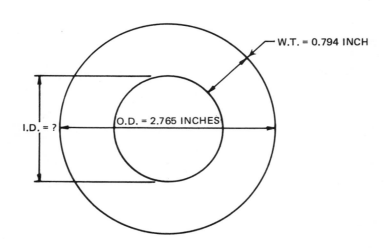

Fig. 11-2

Section 3 — Decimals

3. A typography job set by hand takes 16.7 hours. The same job set by machine is done in 7.25 hours. How much time is saved by using the machine.

4. The thickness of a book shown in Fig. 11-3 is 2.341 inches. The covers are made of stock 0.417 inch in thickness. Determine the thickness of the book before the covers are put on.

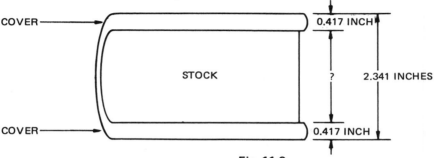

Fig. 11-3

5. A customer is billed $2,345.00 for a printing job. The costs of the labor, materials and overhead are shown:

   Labor $1,240.50
   Materials 712.25
   Overhead........................... 122.00

   Determine the profit left after the labor, materials, and overhead costs are paid.

6. The value of printing paper stock in the storage room is $8,972.50. During one week, $3,250.25 worth of paper is taken out of stock. Determine the value of the remaining stock.

7. The ink storage area of the pressroom has $1,172.50 worth of ink on its shelves. During one month, $575.25 worth of ink is used. Determine the value of the remaining ink.

8. A box of green offset paper costs $126.50. The same size box of white paper costs $97.75. How much more does the box of green paper cost than the box of white paper?

9. The shaft of a cylinder is shown in Fig. 11-4, page 24.
   a. Determine the length of A.
   b. Determine the length of B.
   c. Determine the length of C.

10. An offset printing plant is owned by four persons, Mr. Jones owns 0.35 of the business, Mr. Adams owns 0.15, and Ms. Smith owns 0.45. The balance is owned by Ms. Thompson. How much of the business is owned by Ms. Thompson, if 1.00 is the total of the four persons' shares in the business?

Unit 11 — Subtraction of Decimals

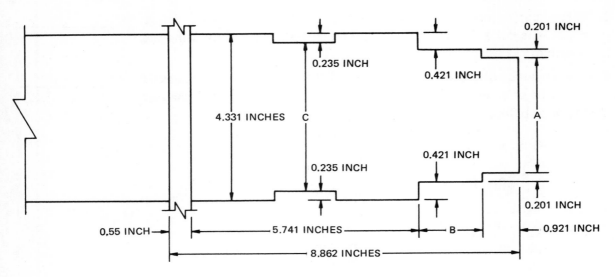

Fig. 11-4

11. A commercial artist is credited with 1.2 hours for Job "A", 3.2 hours for Job "B", and 1.6 hours for Job "C". Determine the amount of time of a seven-hour day not used by the artist on one of the three jobs.

12. The receipts of a printing plant for a month are $26,252.73 and the expenses are $19,367.87. The profit of the plant is the difference between the receipts and the expenses. Determine the profit of the plant for the month.

UNIT 12 MULTIPLICATION OF DECIMALS

BASIC PRINCIPLES OF MULTIPLICATION

- Study Unit 15 in *Basic Mathematics Simplified* for the principles of multiplication as applied to decimals.
- Apply the principles of multiplication to the printing and graphic communications industry by solving the review problems.

REVIEW PROBLEMS

1. How thick are 1,250 sheets of chipboard if each sheet is 0.125 inch thick? _____
2. How thick are 4,750 sheets of index stock if each sheet is 0.009 inch thick? _____
3. A printing plant sells its scrap paper for 32.5¢ per pound. Determine the amount of money the plant receives if it sells 12,275 pounds of scrap paper. _____
4. What is the weight of 629 sheets of bond paper, if one sheet weighs 0.048 pound? _____
5. What is the cost of 72 pounds of offset black ink, if the ink costs $1.19 per pound? _____
6. What is the cost to ship 68 cartons of printed materials if it costs $1.25 to ship one carton? _____
7. Litho film costs $22.25 per box. Determine the cost of 28 boxes of this film. _____
8. One ream of paper costs $29.35. Determine the cost of 9.2 reams of this paper. _____
9. A cameraman works 6.9 hours on Monday, 7.2 hours on Tuesday, and 5.7 hours on Wednesday. The cameraman earns $13.75 for each hour worked. How much does the cameraman earn for the three days? _____
10. During one week, a stripper works 29.7 hours on a certain job. The stripper makes 2.1 flats every hour she works. Determine the number of flats the stripper makes during the week. _____
11. If offset paper costs 37.5 cents a pound, determine the cost of 1.225 pounds of the paper. _____
12. The paper used in the printing of a folder weighs 0.0523 pounds per sheet.
    a. Determine the weight of 6,275 sheets of this paper. _____
    b. The paper costs 39.5 cents per pound. Determine the total cost of the 6,275 sheets of paper. _____
    c. If the cost of the paper is 37.75 cents per pound, determine the cost of the 6,275 sheets of paper. _____
    d. If each piece of paper weighs, 0.04725 pounds, determine the total weight of 6,275 sheets of paper. _____

UNIT 13 DIVISION OF DECIMALS

BASIC PRINCIPLES OF DIVISION

- Study Unit 16 in *Basic Mathematics Simplified* for the principles of division as applied to decimals.
- Apply the principles of division to the printing and graphic communications industry by solving the review problems.

REVIEW PROBLEMS

1. The cost of 42 reams of paper is $478.26. Determine the cost of one ream of paper. _____

2. Thirty-six pounds of ink cost $22.32. Determine the cost of one pound of ink. _____

3. The thickness of thirteen book covers is 1.56 inches. Determine the thickness of one book cover, in inches. _____

4. One ream of paper contains 500 sheets of paper. The weight of one ream of paper is 8.10 pounds. Determine the weight of one sheet of paper, in pounds. _____

5. An apprentice pressman spends 18.2 hours in a class on "Press Maintenance", 13.7 hours in a class on "Feeding and Delivery Systems" and 23.7 hours in a class on "Proper Cylinder Settings."

   a. Determine the amount of time the apprentice spends in the "Press Maintenance" class, in a decimal part of the time spent in the three classes. _____

   b. Determine the amount of time the apprentice spends in the "Feeding and Delivery Systems" class, in a decimal part of the total time spent in the three classes. _____

   c. Determine the amount of time the apprentice spends in the "Proper Cylinder Settings" class, in a decimal part of the time spent in the three classes. _____

6. The book shown in Fig. 13-1 contains 26 signatures and is 2.78 inches thick. Each of the covers is 0.084 inches thick. Determine the thickness of one signature. _____

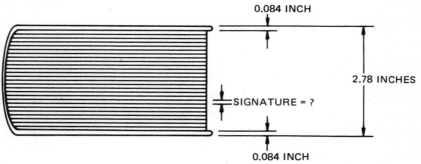

Fig. 13-1

Section 3 — Decimals

7. If a one-pound can of ink costs $2.47, how many one-pound cans can be purchased for $54.34?

8. The cost of maintaining six printing presses for a year is $6,275.00. What is the average cost per week to maintain one press?

9. If twenty-five gallons of the offset roller wash weigh 174.6 pounds, determine the weight of one gallon of wash.

10. If one line of type takes 0.15 hours to set, determine the number of lines of type that are set in 3.7 hours.

11. The cost of 337 inches of litho film is $9.27. Determine the cost of one inch of film.

12. A client's bill is $2,475.70. The client pays the printer one half of the bill in cash and pays the remaining part of the bill in four equal payments. What is the amount of one payment?

13. A printer pays $127.42 for a shipment of brochures at the rate of $.46 a pound. Determine the number of pounds of brochures in the shipment.

14. Change the following fractions to decimals. Round all answers to three decimal places.

    a. 3/8 = _____   h. 25/64 = _____
    b. 3/16 = _____   i. 5/9 = _____
    c. 3/4 = _____   j. 8/9 = _____
    d. 1/2 = _____   k. 3/7 = _____
    e. 5/8 = _____   l. 3/5 = _____
    f. 9/16 = _____   m. 5/12 = _____
    g. 19/32 = _____ n. 7/12 = _____

UNIT 14 READING AN OUTSIDE MICROMETER

BASIC PRINCIPLES OF MICROMETER READING

- Study Unit 19 in *Basic Mathematics Simplified* for the principles of micrometer reading.
- Apply the principles of micrometer reading to the printing and graphic communications industry by solving the review problems.

REVIEW PROBLEMS

Complete the following table to show how the micrometer is set to obtain the given readings.

Problem Number	Reading in Inches	Number of Numbered Divisions On the Barrel	Number of Unnumbered Divisions On the Barrel	Number of Thimble Divisions
1	0.675			
2	0.2135			
3	0.1765			
4	0.333			
5	0.450			
6	0.250			
7	0.009			
8	0.033			
9	0.088			
10	0.575			
11	0.612			
12	0.547			
13	0.912			
14	0.874			
15	0.361			
16	0.745			
17	0.639			
18	0.492			
19	0.268			
20	0.575			

Section 4 — PERCENT

UNIT 15 FRACTIONAL EQUIVALENTS

BASIC PRINCIPLES OF FRACTIONAL EQUIVALENTS

- Study Unit 25 in *Basic Mathematics Simplified* Third Edition for the principles of fractional equivalents as applied to percent; or

  Study Unit 23 in *Basic Mathematics Simplified* **Fourth Edition** for the principles of fractional equivalents as applied to percent.

- To find the fractional equivalent of a percent use the procedure given in the example.

  Example:

  80% = ? fractional equivalent
   Divide 80% by 100 and drop the percent sign
   80% = 80/100

  Reduce the fraction to its lowest terms.

   Divide numerator and denominator by 20
   80/100 ÷ 20/20 = 4/5 (Ans.)

- To find the percent in a problem use the procedure given in the example.

  Example:

  What percent of 80 is 40? Divide 40 by 80
   40/80 = 0.5
    Multiply by 100 and add the percent sign
    0.5 x 100 = 50% (Ans.)

- If the percent and one quantity in a problem are known, the second quantity required can be found.

  Example:

  In a case of 32 bottles, 50% are used. Determine the number of bottles used.

   50% = 50/100 = 1/2
    1/2 x 32 = 16 bottles (Ans.)

- Apply the principles of fractional equivalents to the printing and graphic communications industry by solving the review problems.

REVIEW PROBLEMS

1. Give the fractional equivalent of each of the following percents.

   a. 75% _____   d. 35% _____   g. 66 2/3% _____   j. 37 1/2% ____
   b. 30% _____   e. 40% _____   h. 16 2/3% _____
   c. 60% _____   f. 12 1/2% _____   i. 25% _____

Unit 15 — Fractional Equivalents

2. What percent of 64 is 16? _____

3. What percent of 40 is 5? _____

4. What percent of 175 is 15? _____

5. What percent of 180 is 20? _____

6. What percent of 132 is 6? _____

7. What percent of 36 is 9? _____

8. In a shipment of one gross (144) bottles of developer, 50% of the bottles are found broken. How many broken bottles are in the shipment? _____

9. In a shipment of 400 bottles of hypo, 25% of the bottles are found broken. Determine the number of broken bottles. _____

10. In a shipment of 600 bottles of film cleaner, 10% of the bottles are found broken. Determine the number of broken bottles. _____

11. In a shipment of 80 bottles of gum preservative, 12 1/2% of the bottles are found broken. Determine the number of broken bottles. _____

12. In a shipment of 222 bottles of film stabilizer, 16 2/3% of the bottles are found broken. Determine the number of broken bottles. _____

13. Out of 6,400 printed envelopes, 1,200 are rejected and the rest are accepted.

    a. Determine the percent of envelopes rejected. _____

    b. Determine the percent of envelopes accepted. _____

UNIT 16 SIMPLE PERCENT

BASIC PRINCIPLES OF SIMPLE PERCENT

- Study Unit 26 in *Basic Mathematics Simplified* Third Edition for the principles of simple percent; or
  Study Unit 24 in *Basic Mathematics Simplified* **Fourth Edition** for the principles of simple percent.
- Study Unit 15 of this workbook.
- Apply the principles of simple percent to the printing and graphic communications industry by solving the review problems.

REVIEW PROBLEMS

1. The salary of a pressman is increased from $8,500 a year to $9,100 a year. Determine the percent of the increase in salary.

2. A printing plant buys 3,600 sheets of cover stock. Some of the stock, 15% of the total, is damaged by water; 75% of the total is returned unused; the remainder is used on a job.
   a. Determine the number of sheets damaged by water.
   b. Determine the number of sheets used on the job.
   c. Determine the number of sheets returned unused.

3. A bill for materials is $780.50. Of this amount, 15% is the profit made by the dealer. Determine the profit of the dealer.

4. A bill for letterheads and envelopes amounts to $269.50, including a profit of $40.00. Determine the percent of profit in the bill.

5. The cost of printing a job to a printing company is $1,265.00. The company sells the job for 30% more than their cost to print the job. Determine the selling price of this job.

6. A printing job costs a dealer $325. The dealer adds 20% of the cost when a customer buys the job. Determine the price the customer pays for the job.

7. A job costs a printer $1,210 and is sold to a customer for $1,113.20. Determine the loss to the printer on this job, in percent.

8. A paper merchant sells a quantity of paper at a profit of 12 1/2% and makes a profit of $350.00. What is the cost of the paper?

9. An offset duplicator worth $9,000.00 is sold for 2/3 of its value.
   a. Determine the price at which the duplicator is sold.
   b. The salesman who makes the sale receives 1 3/4% of the price at which the duplicator is sold. How much does the salesman receive?

10. The estimated time to complete a printing job is three hours. The job is completed in 2 1/4 hours. Determine the percent of the estimated time used to complete the job.

Unit 16 — Simple Percentage

11. A salesperson receives 4 1/2% of the value of the paper sold by a company. A company sells $6,260 worth of paper. Determine the amount of money the salesperson receives.

12. During one year, an offset shop income is $51,500. Of this income 42% is used to pay wages, 17% is used to buy materials, 16% is used for new equipment, 2% is used to pay the taxes, 3% is paid for interest on loans, 6% is kept for reserves, 9% is profits made and 5% is used for miscellaneous costs.

   a. Determine the amount of money used to pay wages.

   b. Determine the amount of money used to buy materials.

   c. Determine the amount of money used for new equipment.

   d. Determine the amount of money used to pay taxes.

   e. Determine the amount of money paid for interest on loans.

   f. Determine the amount of money kept for reserves.

   g. Determine the amount of profits made.

   h. Determine the amount of money used for miscellaneous costs.

13. A small printer receives $625.00 for a printing job and makes a profit of 12%. Determine the cost of the job before the profit is added to the cost.

14. A quick copy center sells a printing job for $175.00. The center loses 9% on the job. What is the cost of the job to the copy center.

15. A printer pays $642.50 for the stock, $152.50 for the cold-type composition, and $26.00 for the miscellaneous labor on a job. The printer charges 15% profit on the stock, 10% profit on the composition, and 30% profit on the miscellaneous labor. Determine the total amount of the printer's bill for this job.

16. On a certain job, a printer estimates the cost of the labor will be $455.00, the cost of the stock will be $172.50, and the other miscellaneous expenses will be $40.50. The job requires 95% of the estimated cost of the labor, 90% of the estimated cost of the stock and 110% of the estimated miscellaneous expenses to be spent.

   a. Determine the amount of money spent on the labor needed for this job.

   b. Determine the amount of money spent on the stock needed for this job.

   c. Determine the amount of money spent for the miscellaneous expenses for this job.

   d. Determine the total actual costs of this job.

UNIT 17 DISCOUNTS

BASIC PRINCIPLES OF DISCOUNTS

- Study Unit 30 in *Basic Mathematics Simplified* Third Edition for the principles of discounts; or

  Study Unit 30 in *Basic Mathematics Simplified* **Fourth Edition** for the principles of discounts.

- When discounts are allowed on items purchased there are two prices mentioned in problems. The *net price* is the price before the discounts are determined. The *final price* is the price after the discounts are determined and subtracted from the net price.

  Net Price − Discount(s) = Final price

- Apply the principles of discounts to the printing and graphic communications industry by solving the review problems.

REVIEW PROBLEMS

1. A lithographer's cost of production on a certain job is $1,274.75.
   a. If the lithographer adds 20% to the cost of production for profit, what will the cost of the job be to the customer? _____
   b. The customer is given a discount of 3%. Determine the cost of the job to the customer. _____

2. A lithographer purchases equipment for $2,450.00. The lithographer pays cash and is given a discount of 14 1/2%. Determine the final cost of the equipment. _____

3. When a printer purchases a quantity of ink whose net price is $392.00, a double discount is given. The first discount is 30% and the second discount is 2%. Determine the final price the printer pays for the ink. _____

4. A customer is allowed a regular discount of 25% and a second discount of 2 1/2% on a bill. If the net price of the bill is $1,532.00, determine the final price. _____

5. On a bill of $720.00, two discounts are given. The first discount is 30% and the second discount is 10%. Determine the final price of the bill. _____

6. The net price of a printing bill is $1,500.
   a. If a discount of 35% is given on the bill, determine the final price of the bill. _____
   b. If two discounts, one of 20% and the second of 15%, are given on the bill, determine the final price of the bill. _____
   c. Which final price is less, (a) or (b), and by how much? _____

7. A customer purchases 5,000 letterheads for a price of $6.85 per M, 5,000 envelopes for a price of $4.25 per M, and 2,500 circulars for a price of $49.75. (The abbreviation M means 1000).

33

Unit 17 — Discounts

    a. Determine the total bill for the three items.

    b. The printer gives this customer a discount of 2% if the bill is paid within 30 days. If the customer pays the bill within 30 days, determine the total cost of the three items.

8. The monthly bills owed by a customer are $1,737.62. If the bills are paid on or before the 10th of the next month, the customer is given a 3% discount on the bills. What is the total amount of the bills after the discount is taken?

9. The stock and labor costs for an order of letterheads and envelopes amount to $1,240.60. The printer adds 14% for profit and overhead to the stock and labor costs. If a 3% discount is allowed the customer for cash payment within 10 days, what is the total price?

10. A.B.C. Litho, Inc. purchases a quantity of cover stock for $1,725.00.

    a. The company is given three discounts of 10%, 8%, and 2%. Determine the final price of the cover stock.

    b. If A.B.C. Litho, Inc. is given a flat 20% discount on the paper, will they pay more or less than the price in (a)?

    c. Determine the difference between the prices in (a) and (b).

11. The net price of the parts used to repair an offset press is $67.50, of which $42.00 is subject to a 33 1/3% discount from the net price. What is the total price of the parts?

12. If at least six tubes of black litho ink are purchased, each tube costs $2.65. A discount of 20% is given if the bill for the ink is paid within 10 days. Determine the total bill for twelve tubes of ink if the bill is paid within 10 days.

UNIT 18 PROFIT AND LOSS, COMMISSIONS

BASIC PRINCIPLES OF PROFIT AND LOSS, COMMISSIONS

- Review Unit 26 in *Basic Mathematics Simplified* Third Edition for the principles of percent as applied to profit and loss and commissions; or
  Review Unit 24 in *Basic Mathematics Simplified* **Fourth Edition** for the principles of percent as applied to profit and loss and commissions.
- Commissions are frequently used to pay salespeople. A commission is a certain percentage of the total value of the items a salesperson sells.
  Example: A salesperson receives a commission of 10% of the total value of the paper she sells. She sells $100 worth of paper. Determine the salesperson's commission.
  Commission = 10% = 1/10 = .10
  $100 x .10 = $10 = commission received (Ans.)
- Apply the principles of profit and loss and commissions to the printing and graphic communications industry by solving the review problems.

REVIEW PROBLEMS

1. The total receipts of a printing company for one month are $6,000.00. Of this amount, 40% is for wages, 10% is for rent, 5% is used to pay for heat, and 20% is needed to cover all other miscellaneous overhead expenses. The remaining amount of money is profit.
   a. Determine the amount of profit made during the month. _____
   b. What percent of the total receipts is the profit? _____

2. A bindery buys 1500 feet of stitching wire at a price of 6.5 cents per foot.
   a. Determine the cost of the wire. _____
   b. The following day the cost of the wire increases 12 1/2%. Determine the price of 1500 feet of wire at this increased cost. _____
   c. Determine the difference between the price in (a) and the price in (b). _____

3. A salesperson receives a commission of 5% of the value of the parts he sells. He sells $49.00 worth of parts. What is his commission? _____

4. An agent receives a $112.50 commission for selling one $1,800.00 film developing sink. Determine the commission rate the agent receives. _____

5. A graphic arts camera sells for $4,370.00. A commission of 15% is given to the salesperson for this sale. Determine the amount of commission the salesperson receives. _____

6. An offset duplicator, whose price was $3,655.75, decreases in value 35% in one year. Determine the present value of the duplicator. _____

7. A stripping table costs $1,250.00. The dealer's profit is 25% of the cost of the table. Determine the dealer's profit. _____

Unit 18 — Profit and Loss, Commissions

8. A printing machine normally sells for $1,474.00. The machine's price is reduced 23% for a sale. Determine the sale price of the printing machine. _____

9. A drum of ink contains 54 gallons of ink; each gallon weighs 7 pounds. The price of the ink is 7 1/2 cents per pound.

   a. Determine the cost of the drum of ink. _____

   b. The printer loses 5% of the ink in one drum because of a spill. Determine the value of the ink that is spilled. _____

10. A printer buys a quantity of envelopes for $625.10 and pays $41.40 for the envelopes to be shipped to her printing plant. The printer sells the envelopes so that she makes a profit of 33 1/3% of the total costs of buying and shipping the envelopes. Determine the price the printer receives for the envelopes. _____

UNIT 19 INTEREST AND TAXES

BASIC PRINCIPLES OF INTEREST AND TAXES

- Study Unit 31 in *Basic Mathematics Simplified* Third Edition for the principles of percent as applied to interest and taxes; or
  Study Unit 29 in *Basic Mathematics Simplified* **Fourth Edition** for the principles of percent as applied to interest and taxes.

- Tax rates are commonly expressed in three ways:
  a. mils per dollar of value (1 mil = 1/1000 of a dollar = 1/10 cent = 0.1 cent).
  b. dollars per hundred dollars of value of property or equipment.
  c. Dollars per thousand dollars of value of property or equipment.

  These tax rates can be changed to percents.
  Example: What percent does a tax rate of $8.50 per hundred dollars equal?
  $8.50/$100 = 8.5% (Ans.)

- Interest is determined in the same way discounts and profit and commissions (Units 16, 17, 18 in this workbook) are determined. The general formula to determine interest on savings accounts is shown:

  interest earned =
  $$\frac{\text{money in account} \times \text{annual interest rate} \times \text{months money is in account}}{12}$$

  For money left in a savings account for more than one year, interest is received on the money in the account (the principal) plus on the interest earned during the first year.
  Example: There is $500 in a savings account. The bank gives 5% interest per year. Determine the amount of money in the account after two years.

  $500 x .05 = $25 + $500 = $525 after one year
  $525 x .05 = $26.25 + $525 = $551.25 (Ans.)

- Banks charge interest on money borrowed by their customers. The interest due on a loan is determined in the same way interest is determined. Some banks make a type of loan with the *interest discounted*. This means the interest due is determined and deducted from the amount being borrowed.
  Example: A customer borrows $100 from a bank at 8% interest discounted. The interest due is determined as shown:

  $100 x .08 = $8

  This amount is deducted from $100. Thus, the customer receives $92 but has to pay back $100. The actual interest rate is $8/$92 = 8.70%.

- Apply the principles of interest and taxes to the printing and graphic communications industry by solving the review problems.

REVIEW PROBLEMS

1. What percent does a tax rate of 30 mils per dollar equal? _____

2. What percent does a tax rate of $4.50 per hundred dollars equal? _____

3. What percent does a tax rate of $32.50 per thousand dollars equal? _____

Unit 19 — Interest and Taxes

4. Determine the interest earned if $800 is left in a savings account for one year. The interest rate is 6% per year. _____

5. A lithographer keeps $4,200 in a savings account for two years. Determine the amount of money in the account after two years if the interest rate is 5 1/2% per year. _____

6. A printing plant pays tax on the equipment in the plant. The tax rate is 30 mils per dollar of value of the equipment. The value of the equipment is $15,000.00. What is the tax bill for the plant? _____

7. A printer deposits $1,745.00 in a savings account for one year. If the annual interest rate is 5 1/2%, determine the interest the printer earns on the money in the account. _____

8. A pressman borrows $1,000 from a bank for one year at an interest rate of 8 1/2% per year.

   a. Determine the amount of money the pressman pays back to the bank. _____

   b. The pressman pays back the money in 10 equal payments. Determine the amount of each payment. _____

9. A printer pays $866 interest to a bank for money he borrowed for one year at an interest rate of 5 3/4%. How much did the printer borrow from the bank? _____

10. A person deposits $50 in a savings account each month for twelve months. The interest earned for the twelve months is $6.06. Determine the interest rate given by the bank. _____

11. The interest due on a printer's capital investment is a cost of doing business. A printer is charged 9% interest on the printing equipment purchased for the plant. The press equipment in the plant is worth $14,460, the composing equipment is worth $6,040 and the bindery is worth $1,242. Determine the interest due on the printer's equipment. _____

12. The federal income tax on payroll earnings is 22%. Determine the tax withheld on a paycheck of $187.25. _____

13. A person borrows $600 with the interest discounted. The interest rate is 8% per year.

   a. Determine the interest due on the money borrowed. _____

   b. What is the actual rate of interest? _____

14. A person borrows $8,735 with the interest discounted. The interest rate is 9.75%. How much money does this person actually receive from the bank? _____

SECTION 5 — MEASUREMENT

UNIT 20 LINEAR MEASURE

BASIC PRINCIPLES OF LINEAR MEASURE

- Study Units 18 and 19 in *Basic Mathematics Simplified* Third Edition for the principles of linear measure; or
  Study Unit 18 in *Basic Mathematics Simplified* **Fourth Edition** for the principles of linear measure.
- Carry all answers to two decimal places.
- Apply the principles of linear measure to the printing and graphic communications industry by solving the review problems.

REVIEW PROBLEMS

1. Change 14.4 miles to feet. _____
2. Change 6.75 miles to feet. _____
3. Change 3.25 miles to inches. _____
4. Change 5 feet 3 inches completely to feet. _____
5. Change 3 feet 2 inches completely to feet. _____
6. Change 9 feet 8 inches completely to feet. _____
7. How many feet are in one mile? _____
8. How many yards are in one mile? _____
9. Change 3 1/2 miles to yards. _____
10. Change 1 1/4 miles to yards. _____
11. What percent of a mile is 1,000 feet? _____
12. Change 1,250 inches to yards. _____
13. Change the following lengths to inches:
    a. 6 feet _____   c. 3 feet 2 1/4 inches _____
    b. 1 foot 6 inches _____   d. 2 feet 3 7/16 inches _____
14. Change the following lengths to feet and inches.
    a. 9.9 feet _____   c. 3.65 feet _____
    b. 13.62 feet _____   d. 13.19 feet _____
15. Express the following lengths in feet as a decimal.
    a. 13 feet 6 inches _____   c. 4 feet 2 1/4 inches _____
    b. 8 feet 5 3/4 inches _____   d. 5 feet 4 1/2 inches _____

Unit 20 – Linear Measure

16. An offset printing press frame measures 12 feet 6 5/8 inches. What is the total length of the frame, in inches?

17. Three pieces of stitching wire are cut from a roll which is 20 feet long. The lengths of the pieces are 2 feet 4 1/2 inches, 3 feet 11 1/4 inches and 4 feet 9 3/8 inches. Determine the length of wire remaining on the roll, in feet and inches.

18. Two pieces of plastic tubing, each measuring 2 feet 3 7/8 inches, are cut from a coil 10 feet long. Determine the length of tubing remaining on the coil, in feet and inches.

19. From a roll of litho film 50 feet long, 12 pieces of film, each 14 inches long, are cut. Determine the length of film remaining on the roll, in feet.

20. A photographer wishes to put three booths against the wall shown in Fig. 20-1. Each booth is 4 feet 6 3/4 inches wide. How much wall space will be taken up if the three booths are placed side by side?

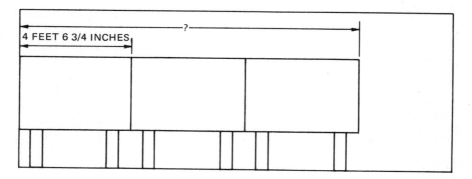

Fig. 20-1

21. The length of the parking lot shown in Fig. 20-2 is 130 feet. An opening of ten feet is allowed for the roadway, and no cars are parked in this area. If each car takes up 7 feet 6 inches, how many cars can be parked along the indicated wall in the parking area?

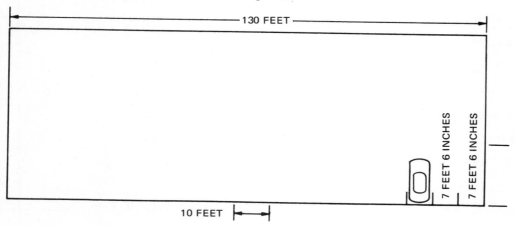

Fig. 20-2

UNIT 21 METRIC SYSTEM

BASIC PRINCIPLES OF THE METRIC SYSTEM

- Study Unit 23 in *Basic Mathematics Simplified* Third Edition for the principles of measurement as applied to the metric system; or
Study Unit 36 in *Basic Mathematics Simplified* **Fourth Edition** for the principles of measurement as applied to the metric system.
- Study the table of Metric Linear Measure, Fig. 21-1.

Linear Measure

10 millimeters (mm) = 1 centimeter (cm)
10 centimeters (cm) = 1 decimeter (dm)
10 decimeters (dm) = 1 meter (m)
10 meters (m) = 1 dekameter (dam)
10 dekameters (dam) = 1 hectometer (hm)
10 hectometers (hm) = 1 kilometer (km)

Fig. 21-1

- Apply the principles of the metric system of measurement to the printing and graphic communications industry by solving the review problems.

REVIEW PROBLEMS

1. Determine the number of mm in fifteen dm.   _____
2. Determine the number of cm in twenty dm.   _____
3. Determine the number of dm in thirty hm.   _____
4. Change 14.387 km to hm.   _____
5. Change 14,307 cm to m.   _____

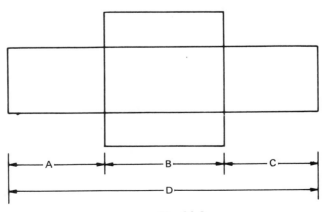

Fig. 21-2

Unit 21 — Metric System

6. Fig. 21-2 shows a part of a printing machine.
   a. If length A is 55 mm, length B is 40 mm, and length C is 50 mm, determine length D, in mm.
   b. If length D is 6 m, length B is 14 dm, and length C is 20 dm, determine length A, in dm.
   c. If length B is 18.5 cm, length C is 1.45 dm, and length D is 4.92 dm, determine length A, in cm.

7. Fig. 21-3 shows three rooms in a printing plant.
   a. Determine the area of room A, in m².
   b. Determine the area of room C, in m².
   c. Determine the total area of rooms A and B, in m².

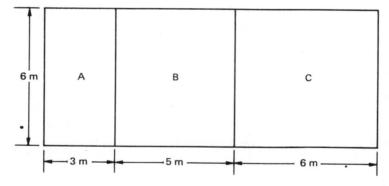

Fig. 21-3

8. Fig. 21-4 shows a piece of paper 15 m long. Determine the number of pieces of paper that can be cut from this piece, if each piece is 2.5 cm long.

Fig. 21-4

UNIT 22 METRIC EQUIVALENTS

BASIC PRINCIPLES OF METRIC EQUIVALENTS

- Study Unit 23 in *Basic Mathematics Simplified* Third Edition for the principles of measurement as applied to metric equivalents; or
  Study Unit 36 in *Basic Mathematics Simplified* **Fourth Edition** for the principles of measurement as applied to metric equivalents.
- Apply the principles of the metric system of measurement to the printing and graphic communications industry by solving the review problems.

REVIEW PROBLEMS

1. Determine the number of inches in 2 dm. _____
2. Determine the number of feet in 30 hm. _____
3. Determine the number of cm in 25 feet. _____
4. The length of a piece of paper is 8.5 inches. Determine its length in mm. _____
5. How many pieces of paper, each 1 m long, can be cut from a length of paper 42 feet long? _____
6. Fig. 22-1 shows the measurement of a printing room.
   a. Change the length to feet. _____
   b. Change the width to yards. _____
   c. Determine the area of the room in square yards. _____
7. A package of 500 sheets of a certain paper is 2.11 inches thick. Determine the thickness of each sheet of paper, in mm. _____
8. A printing plate is 14 inches long. Determine the length of the plate in cm. _____

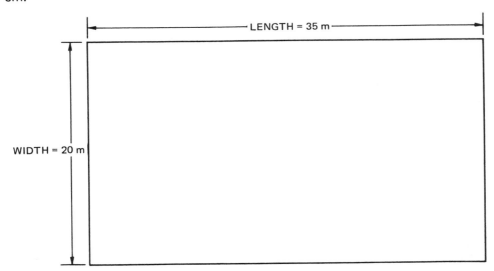

Fig. 22-1

UNIT 23 ANGULAR MEASUREMENT

BASIC PRINCIPLES OF ANGULAR MEASUREMENT

- Study Unit 20 in *Basic Mathematics Simplified* Third Edition for the principles of angular measurement; or

  Study Unit 19 in *Basic Mathematics Simplified* **Fourth Edition** for the principles of angular measurement.

- If the sum of two angles, A and B, is 90°, angle A is said to be the complement of angle B, and angle B is said to be the complement of angle A. In Fig. 23-1, angle A and B are complementary angles.

  angle A + angle B = 90°

- If the sum of two angles, C and D, is 180°, angle C is said to be the supplement of angle D and angle D is said to be the supplement of angle C. In Fig. 23-1, angle C and angle D are supplementary angles.

  angle C + angle D = 180°

- Apply the principles of angular measurement to the printing and graphic communications industry by solving the review problems.

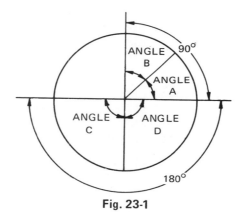

Fig. 23-1

REVIEW PROBLEMS

1. Determine the number of degrees in a right angle.
2. Determine the number of degrees in a straight angle.
3. Determine the number of 60° angles in a full circle.
4. Determine the number of 45° angles in a straight angle.
5. Determine the number of 30° angles in a full circle.
6. If angle A and angle B are complementary angles and angle A is 38°, determine angle B.
7. If angle A and angle D are complementary angles and angle D is 48° 42', determine angle A.
8. If angle C and angle E are supplementary angles and angle C is 135° 40', determine angle E.

9. Fig. 23-2 shows complementary angles A and B. Determine angle B. _____

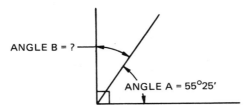

Fig. 23-2

10. Fig. 23-3 shows supplementary angles C and D. Determine angle C. _____

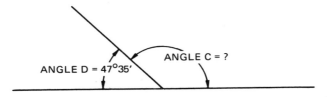

Fig. 23-3

UNIT 24 UNITS OF AREA AND VOLUME MEASURE

BASIC PRINCIPLES OF AREA AND VOLUME

- Study Units 21 and 22 in *Basic Mathematics Simplified* **Third Edition** for the principles of area and volume; or
  Study Units 20 and 21 in *Basic Mathematics Simplified* **Fourth Edition** for the principles of area and volume.
- Study the Specific Gravity Chart, Fig. 24-1.

Substance	Specific Gravity
Water	1.000
Sulfuric Acid	1.835
Alcohol	0.816

Fig. 24-1

Specific gravity refers to the weight of a given amount of a substance compared to the weight of the same amount of water. Sulfuric acid is 1.835 times as heavy as water in equal amounts.

- Study Fig. 24-2.

```
1 cubic foot (cu. ft.) = 1,728 cubic inches (cu. in.)
1 U.S. gallon = 231 cubic inches (cu. in.)
1 cubic foot of water weighs 62.5 pounds (lbs.)
1 U.S. gallon of water weighs 8.333 pounds (lbs.)
1 Imperial gallon = 277.274 cubic inches (cu. in.)
43,560 square feet (sq. ft.) = 1 acre
```

Fig. 24-2

- Carry all answers to three decimal places.
- Apply the principles of area and volume to the printing and graphic communications industry by solving the review problems.

REVIEW PROBLEMS

1. a. How many square feet are there in a parking area 200 feet long and 125 feet wide? _____

   b. Determine the number of acres in the parking area. _____

2. Rubber material for flexographic printing plates is purchased by the square foot.

   a. If a roll of the rubber material is 20 inches wide, what length of the material is needed to obtain 25 square feet? _____

   b. Determine the number of printing plates 1 1/2 feet wide and 14 inches long that can be cut from the piece of rubber material in (a). _____

   c. Determine the area of one printing plate, in square inches. _____

Section 5 — Measurement

3. Fiberglass costs $3.45 per square foot. What is the cost of a piece 4 feet long and 42 inches wide? _____

4. A stockroom used for storing solvents is to be fireproofed. The floor and ceiling are 10 feet long and 8 feet wide and the four sides are 8 feet wide and 7 feet high. Determine the total area of the ceiling, the floor and the four sides of the stockroom. _____

5. A space 12 feet long and 8 feet wide is allowed for one offset press in a storage building. How many presses can be stored in a building with an area of 4,000 square feet? _____

6. How many square inches are there in a case of zinc plates used for photoengraving if each plate is 14 inches long and 20 inches wide and there are 144 plates in the case? _____

7. The owner of a store, 40 feet long and 30 feet wide, charges $1.25 per square foot per month rental. What is the rent per month? _____

8. How many square inches are there in a package of golden rod material used for stripping if each sheet is 17 inches long and 22 inches wide and there are 750 sheets in a package? _____

9. What is the weight of one gallon of alcohol? _____

10. How many pints does one pound of alcohol contain? _____

11. What is the weight of 50 gallons of alcohol? _____

12. The legal load limit of a 1-ton delivery truck is 1,800 pounds. A 550-gallon tank weighing 185 pounds is mounted on a 1-ton truck. The tank is filled with water. By how many pounds does the load on the truck exceed the legal limit? _____

13. a. How many U.S. gallons are there in one cubic foot of water? _____

    b. How many Imperial gallons are there in one cubic foot of water? _____

14. Determine the number of U.S. pints in one Imperial gallon. _____

15. Determine the number of U.S. gallons in ten Imperial gallons. _____

16. How many Imperial gallons can a 15-gallon (U.S.) tank hold without overflowing? _____

47

UNIT 25 TIME AND MONEY CALCULATIONS

BASIC PRINCIPLES OF TIME AND SPEED

- Study Unit 29 in *Basic Mathematics Simplified* for the principles of time and speed.
- Apply the principles of time and speed to the printing and graphic communications industry by solving the review problems.

REVIEW PROBLEMS

1. A time card shows that a worker starts a stripping job at 8:27 A.M. and finishes at 3:40 P.M. The worker earns $4.65 per hour. Allowing an hour for lunch, how much does the worker earn for the work?

2. The flat rate price for the labor on a job is $49.00. The job takes 5 1/2 hours. What is the hourly wage for the labor on the job?

3. If the time allowance for a certain stripping job is twenty minutes, how many jobs can be done in eight hours?

4. A mechanic is paid $3.75 per hour. How many hours are required to complete a job when the allowance for labor is $81.25?

5. A printer works eight hours a day. How many days does he work to complete a job requiring 112 hours?

6. A photo job takes 240 hours to complete. A photographer works 35 hours each week until the job is completed. How many weeks does the photographer work?

7. How many minutes are in six hours?

8. If a worker is 28 minutes late, how many tenths of an hour are deducted from his time?

9. A photographer earns $21.00 for four hours of work. Determine the hourly wage of the photographer.

10. A plant manager starts work at 8:30 A.M. and finishes for the day at 5:15 P.M. She takes 45 minutes for lunch and 10 minutes for each of two coffee breaks. Determine the total working time of the plant manager.

Section 6 — GRAPHS

UNIT 26 PRACTICAL APPLICATIONS OF GRAPHS AND CHARTS

BASIC PRINCIPLES OF GRAPHS AND CHARTS

- Study Units 33, 34, and 35 in *Basic Mathematics Simplified* Third Edition for the principles relating to graphs and charts; or
  Study Units 31, 32, and 33 in *Basic Mathematics Simplified* **Fourth Edition** for the principles relating to graphs and charts.
- Apply the principles of graphs and charts to the printing and graphic communications industry by solving the review problems.

REVIEW PROBLEMS

1. a. Make a graph by plotting units of paper used against weeks for a ten-week period using the following figures:

Week	1	2	3	4	5	6	7	8	9	10
Units of paper used	80	70	70	60	70	75	90	110	100	90

   b. Find the weekly average of units of paper used during the ten weeks.　　_____

   c. Determine the number of weeks during which an above the weekly　　_____
      average amount of paper was used.

   d. Determine the number of weeks that a below the weekly average　　_____
      amount of paper was used.

2. Answer the following questions based on Fig. 26-1.

   a. During which month was the largest quantity of paper used?　　_____

   b. During which month was the smallest quantity of paper used?　　_____

   c. Determine the monthly average of paper used at this printing plant.　　_____

   d. How many pounds of paper are used in January?　　_____

   e. How many pounds of paper are used in March?　　_____

   f. How many pounds of paper are used in June?　　_____

   g. How many pounds of paper are used in August?　　_____

   h. How many pounds of paper are used in November?　　_____

   i. Did the printing plant use more or less than the monthly average　　_____
      during September?

Unit 26 — Practical Applications of Graphs and Charts

j. Did the printing plant use more or less than the monthly average during December? _____

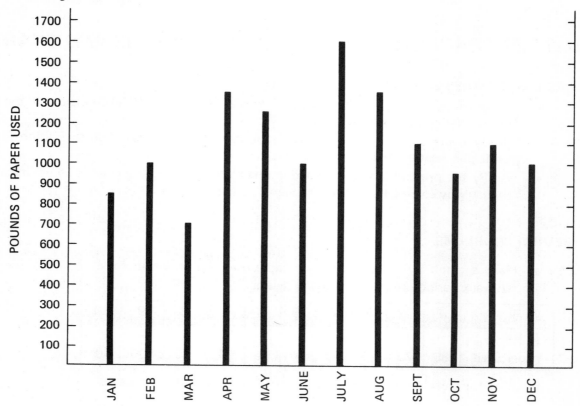

Fig. 26-1

3. Fig. 26-2 shows the weekly income of a printer.
   a. How much did the printer earn during the third week? _____
   b. How much did the printer earn during the tenth week? _____
   c. How much did the printer earn during the eighth week? _____

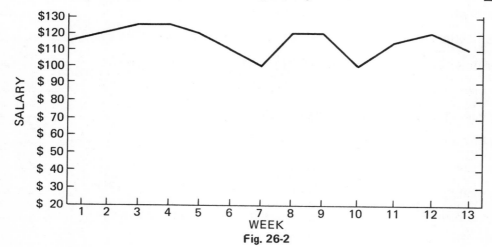

Fig. 26-2

Section 6 — Graphs

4. Make a circle graph illustrating the distribution of total expenses of a printing company for one year as follows:

Wages	35%
Materials	30%
Replacement of machinery	15%
Light and power	5%
Dividends	5%
Interest and taxes	10%

5. Make a circle graph of the budget of an apprentice cameraman who earns $800 each month and spends the money as follows:

Food	$200
Rent and utilities	$200
Clothes	$ 75
Savings	$ 80
Insurance	$ 50
Car Payments	$ 50
Miscellaneous	$145

6. Plot a bar graph for the monthly output of an offset printing plant for the following year:

Month	Output
Jan.	450
Feb.	510
Mar.	300
April	680
May	960
June	780
July	900
Aug.	460
Sept.	685
Oct.	210
Nov.	305
Dec.	115

Unit 26 — Practical Applications of Graphs and Charts

7. Plot a line graph for the monthly output of an offset printing plant for a two-year period. (Use contrasting lines for each year and use the same graph for both lines.)

First year	Month	Output
	Jan.	670
	Feb.	325
	March	770
	April	800
	May	1075
	June	925
	July	930
	Aug.	745
	Sept.	620
	Oct.	450
	Nov.	340
	Dec.	365

Second year	Month	Output
	Jan.	815
	Feb.	625
	March	710
	Apr.	790
	May	900
	June	820
	July	770
	Aug.	810
	Sept.	920
	Oct.	750
	Nov.	610
	Dec.	525

SECTION 7 — PAPER STOCK

UNIT 27 PACKAGING PAPER STOCK

BASIC PRINCIPLES OF PACKAGING PAPER STOCK

- The paper used in the printing industry is produced in many thicknesses and widths, ranging from twelve inches to fourteen feet. Manufacturers of papermaking machines are redesigning these machines to accommodate the larger paper widths. As the finished paper comes off the production machine, it is slit into standard widths with the use of cutting devices. These standard width rolls are then cut into standard-sized sheets; the sheets are gathered, counted, and packaged in quantities of **100-500 sheets**. Fig. 27-1 indicates standards used in packaging paper for printers' use.

```
                    PRINTERS' PAPER TABLE

  1 full (sealed) package of paper stock    = 500 sheets (D)
  1 full box of commercial envelopes        = 500 envelopes
  1 full (sealed) package of lightweight cover    = 250 sheets
  1 full (sealed) package of lightweight blotting = 250 sheets
  1 full (sealed) package of heavyweight blotting = 125 sheets
  1 full (sealed) package of card stock, blanks, bristols, heavy-
      weight and specialty covers = 100 sheets (C)
  1 carton = 1 printer's bundle = approx. 125 lbs.
  4 cartons (bundles) = 1 case = approx. 500 lbs.
  4 cartons (4 bundles or 1 case) = 1 shipping unit
  16 cartons = 1 ton = approx. 2000 lbs.
```

Fig. 27-1

- The quantity of sheets of paper in a package or carton is usually indicated by Roman numerals: M = 1000 sheets, D = 500 sheets, and C = 100 sheets. Paper is usually sold on the basis of **1000 sheets**. The weight of certain paper is indicated by a substance number (in pounds) and/or the actual weight of one ream (500 sheets) of the paper. This weight is printed on the package end flap. To obtain the weight of **1000** sheets of a particular paper, the substance number is doubled.

- Most printer's paper is wrapped and sealed in packages weighing about **125 pounds**. These packages are then shipped in cardboard cartons. Paper mills do not ship broken (less than full) cartons; only a local paper vendor will attempt this practice.

Unit 27 — Packaging Paper Stock

- Study Figs. 27-2, 27-3, 27-4, and 27-5 to become familiar with standard packing schedules.
- Apply the principles of packaging paper stock to the printing and graphic communications industry by solving the review problems.

Packing Schedule of Bristol Boards

Index Bristols

Size	Approx. Caliper.	Weight	Sheets to Cnt.	Sheets to Case
20½ x 24¾	.007½	117	1200	4800
22½ x 28½	.007½	148	1000	4000
25½ x 30½	.007½	180	800	3200
20½ x 24¾	.008½	144	1000	4000
22½ x 28½	.008½	182	800	3200
25½ x 30½	.008½	220	600	2500
20½ x 24¾	.011	182	800	3200
22½ x 28½	.011	230	600	2600
25½ x 30½	.011	280	500	2000
20½ x 24¾	.013½	222	600	2500
22½ x 28½	.013½	280	500	2000
25½ x 30½	.013½	340	400	1700

Printing (Mill)

Ply	Weight per M	Sheets to Ctn.	Sheets to Case
1	180	800	3200
2	200	700	2800
3	240	600	2400
4	280	500	2000
6	320	400	1600

Fig. 27-2

Carton Packing Schedule for Cover Papers

Size	Wgt. per Rm.	Sheets per Pkg.	Sheets per Ctn.
20 x 26	40	500	2000
	50	500	1500
	60	250	1000
	65	250	1000
	80	100	800
	130	100	500
23 x 35	62	500	1000
	77½	250	1000
	93	250	750
	100½	250	750
	124	250	500
	201	100	300
26 x 40	80	250	1000
	100	250	750
	120	125	500
	130	250	500
	160	250	500
	260	100	300

Fig. 27-3

APPROXIMATE CASE CONTENTS OF BOOK PAPERS

1000 Sheet Weight	Sheets per Case
78	7500
80	7500
83	7000
90	6500
99	6000
100	6000
116	5000
120	5000
130	4500
140	4250
148	4000
156	3750
160	3750
166	3500
178	3250
182	3250
198	3000
200	3000
208	2750
222	2500
232	2500
238	2500
240	2500
256	2250
266	2250
280	2000
320	1750
332	1750

Fig. 27-4

Carton Packing of Envelopes

Sizes	Cart. Quant.	Sizes	Cart. Quant.
5	5M	9	5M
6¼	5M	10	2½ M
6¾	5M	11	2½ M
7	5M	12	2½ M
Monarch	5M	14	2½ M

Fig. 27-5

Section 7 — Paper Stock

REVIEW PROBLEMS

1. An order calls for 1,500 sheets of Bond paper. Determine the number of packages of paper in this order. _____

2. An order calls for 9,000 number 10 envelopes.
   a. Determine the number of boxes of envelopes in the order. _____
   b. Determine the number of cartons in the order. _____

3. An order calls for 1,750 sheets each of lightweight and heavyweight blotting.
   a. Determine the number of packages in the order for lightweight. _____
   b. Determine the number of packages in the order for heavyweight. _____

4. An order calls for 6,500 sheets of card stock. How many packages are required to supply this amount? _____

5. An order calls for 3,500 sheets of 20" x 26" cover paper (50-lb. weight).
   a. How many full packages will be shipped? _____
   b. How many cartons will be shipped? _____

6. An order calls for 7,000 sheets of 20" x 26" cover paper (50-lb. weight).
   a. How many full packages can be shipped? _____
   b. How many cartons can be shipped? _____
   c. Is the order large enough for a case? _____

7. An order calls for 3,500 sheets of cover paper (substance 80, 20" x 26").
   a. How many full packages can be shipped? _____
   b. How many cartons are contained in the order? _____
   c. Is the order large enough for a case? _____

8. An order calls for 9,500 sheets of heavyweight blotting.
   a. How many full packages can be shipped? _____
   b. Is there enough for a carton? _____
   c. Is there enough for a case? _____

9. How many cartons are contained in
   a. a case? _____
   b. a ton? _____

10. a. What is the *selling unit* for paper? _____
    b. What is the *shipping unit* for paper? _____

UNIT 28 BASIC SIZE, THICKNESS, AND WEIGHT OF STOCK

BASIC PRINCIPLES OF STOCK SIZE, THICKNESS, AND WEIGHT

- Paper mills manufacture many varieties of paper and paper stock which differ in size, weight, color and composition. This paper stock is usually ordered with the intended use in mind. All paper is classified into categories with a direct relationship to the basic sizes and substance numbers. These categories will assist in designating paper thickness and the weight of 500 sheets (one ream).

  Fig. 28-1 is a table of the basic sizes in which the various types of paper and paper stock are produced.

TABLE OF BASIC SIZES			
Bond and Ledgers Writings, Mimeograph, Duplicator, Gummed, Flats	17 x 22	Card Stock Rope, Bogus, Tag, Folding, Postcard Ticket, Wedding	22½ x 28½
Blotting papers Plain, Coated	19 x 24	Cover papers Specialty covers	20 x 26
Blanks Translucents, Railroad, Tough check, Poster, Calendar, Signboard, Playing Card	22 x 28	Manilas Poster, Kraft	24 x 36
		Newsprint Poster paper	24 x 36
Board Binders, Cloth Corrugated	20 x 30	Padding board Strawboard, Chipboard, Boxboard	26 x 38
Book papers Offset, Label, Text, Ballot, Coated, Uncoated	25 x 38	Pressboard Tagboard	24 x 36
		*Thin papers Manifold, Onion skin Tissues, Parchment	20 x 30
Bristols Mill Index	20½ x 24¾	Tympan papers (.006) in rolls 15"-70" widths and cut in sizes for most presses	24 x 36
Carbon papers Pencil, Typewriter, Copying (Hectograph) One side, both sides	22 x 34	*Wrappings Manila, White Specialty	24 x 36

*In the category "Wrappings", the material comes in sheets or with rolls. There are 480 sheets to a package and rolls of various widths and weights are available. Other than standard sizes are available, but are considered to be special orders.

Fig. 28-1

Section 7 — Paper Stock

- The thickness of paper is designated by one ream (500 sheets) of the basic size for that paper. Sizes other than the basic size vary according to the area in square inches involved. Whenever paper has the same description and the same thickness, it is considered to have the same substance number.

38" 60 LBS.
25"
25 x 38 – 60 equals
60 lbs. for 500 sheets

60 LBS.
25"
38 x 50 – 60 equals
120 lbs. for 500 sheets

25 x 38 – 60 is read as 25 x 38 SUBSTANCE 60, or as 25 x 38 BASIS 60, and may be written as:
25 x 38 – 60
25 x 38 – Sub. 60
25 x 38 – 60 D
25 x 38 – 120 M (S.60)
25 x 38 – 120 M
25 x 38 Basis 60

Fig. 28-2

- The basic size and substance numbers are usually listed in paper dealer catalogs and price lists, as shown in Fig. 28-3. Often there are additional listings of other sizes of paper having the weight of 1000 sheets of the same thickness.

BOOK PAPER

Trade Name of Paper			
25 x 38	60	80	100
25 x 38	120	160	200
28 x 42	148	198	248
35 x 45	198	266	332

Fig. 28-3

- The grain of a certain type of paper is given on the end label of the paper stock package. An end label is shown in Fig. 28-4. The number 22 indicates the direction of the grain and is underlined. This label indicates that the grain is running the long way on each sheet of paper. The direction of the grain can also be indicated as grain short or grain long.

17 x 22 – 16 Substance 16 D 16 lbs. M 32 lbs.	NAME BRAND BOND PAPER	WHITE One Ream 500 Sheets

Fig. 28-4

- The number 16, (17 x 22 – 16) in Fig. 28-4, indicates the substance weight of one ream of this paper. The weights of 500 sheets of this paper (D-16 lbs.) and 1,000 sheets (M-32 lbs.) are also given.

- The standard unit of thickness for measuring paper or board is the *ply*. A ply is measured in points; each point equals 0.001 inch. Thus, 50-point paper is 0.050 inch thick.

Unit 28 — Basic Size, Thickness, and Weight of Stock

- Coded board, or blanks, is also sold by the ply or layer and is manufactured in point thicknesses. Each layer or ply represents approximately 0.003 inch or 3 points. As the number of plys increases, the measurement in points also increases.
- Heavier boards, such as those used for the pad backing of books, are wrapped in packages of fifty pounds each, with the number of sheets in each package varying according to the thickness of the sheet.

To find the number of sheets in one package of binders board, multiply the area of the sheet in square inches by the thickness of the sheet in thousandths of an inch and divide the product into 1456 points.

FORMULA:
$$\frac{1456 \text{ points}}{\text{Sheet area in square inches} \times \text{thickness in thousandths of an inch}} = \text{sheets in 1 package}$$

Example: Find the number of sheets of 26" x 32" — 70-point binders board in one bundle.

① Multiply the area of the sheet by thickness of sheet in thousandths	① 26 x 32 = 832 square inches 832 x .070 = 58.24 points
② Divide the product into 1456 points	② 1456 ÷ 58.24 = 25 sheets in 1 bundle

26 x 32 x .070 = 58.24)1456.00 = 25 sheets

- Paper manufacturers apply standard tests to check the thickness of their paper during the manufacturing process. Special micrometers (which read in thousandths of an inch) are used to measure the thickness. The thickness tests are made at frequent intervals during the manufacturing process. Paper manufacturers can maintain consistency in paper thickness and quality by the use of automatic equipment and good quality control systems.
- Apply the principles of stock size, thickness, and weight to the printing and graphic communications industry by solving the review problems.

REVIEW PROBLEMS

1. What is the basic size for bond and writing paper? _____
2. What is the basic size for book paper? _____
3. What is the basic size for cover paper? _____
4. What is the basic size for card stock? _____
5. How is grain indicated on a package label? _____
6. a. What is the *standard* unit of thickness? _____
   b. In what unit is thickness measured? _____
7. What will 1000 sheets of book paper, substance 70, weigh? _____
8. If 2 plys of card stock are 0.012 inch thick, and each additional ply is 0.003 inch, determine the number of points in 5 plys. _____

UNIT 29 EQUIVALENT WEIGHTS OF STOCK

BASIC PRINCIPLES OF STOCK WEIGHTS

- Whenever paper dealers' catalogs or price lists are unavailable, the weight of paper stock may be determined by one of two methods: A. the square inch method or B. the paper dealers method.

A. Square Inch Method

To find the equivalent weight of paper stock in a size other than the basic size, multiply the area of the desired sheet by the basic weight and divide by the area of the basic-size sheet.

FORMULA: Equivalent Weight = $\dfrac{\text{square inches of desired sheet} \times \text{basic weight}}{\text{square inches of the basic-size sheet}}$

Example: What is the equivalent weight of 17" x 22" — 20 Bond paper in 17 x 28" size.

①	Find the square inches of the desired sheet	①	17 x 28 = 476 sq. in.
②	Find the square inches of the basic size sheet	②	17 x 22 = 374 sq. in.
③	Multiply square inches of desired sheet by the basic weight	③	476 x 20 = 9520
④	Divide by square inches of basic size sheet	④	9520 ÷ 374 = 25.4 + or 25½ lbs. (for 500 sheets) Ans.

$$\dfrac{17 \times 28}{17 \times 22} \times \dfrac{20}{1} = \dfrac{\cancel{17} \times \cancel{28}^{14} \times 20}{\cancel{17} \times \cancel{22}_{11}} = \dfrac{280}{11} = 25.4 + \text{ or } 25½ \text{ lbs. Ans.}$$
(for 500 sheets)

The equivalent weight can also be determined by proportion, either arithmetically or algebraically.

B. The Paper Dealers Factoring Method

To find the equivalent weight of paper stock in a size other than basic size, multiply the dimensions of the desired sheet by the weight factor for that substance number (basic weight).

FORMULA: Equivalent Weight = $\begin{array}{c}\text{Inch dimensions} \\ \text{of desired sheet}\end{array}$ x $\begin{array}{c}\text{weight factor} \\ \text{for basic weight}\end{array}$

Example: What is the equivalent weight of 17" x 22" — 20 Bond paper in 17" x 28" size?

①	Determine the square inches of desired sheet	①	17 x 28 = 476 sq. in.
②	Refer to the decimal weight factor table to find multiplying factor for basic weight of paper	②	Table indicates multiplying factor for substance 20 to be .107 (See Fig. 29-1, page 60).
③	Multiply to obtain weight of 1000 sheets of paper	③	476 x .107 = 50.932 or 51 lbs. (for 1000 sheets) Ans.

17 x 28 x .107 = 476 x .107 = 50.932 or 51 lbs. (for 1000 sheets) Ans.

NOTE: Raise a decimal fraction of a pound to the next half.

Unit 29 — Equivalent Weights of Stock

- The multiplying factor may also be calculated directly by the following method:

> To find the multiplying factor for the basic weights for each of the 17 basic sizes of paper stock, divide the square inches of the basic-size sheet into the basic weight doubled.

FORMULA: $\dfrac{\text{basic weight} \times 2}{\text{basic size (sq. in.)}}$ = weight of 1 sq. inch = factor

Example: What is the decimal weight factor for 1,000 sheets of Bond paper (17" x 22") having a basic weight of 20 pounds in a size other than the basic size?

①	Determine the basic weight of paper	①	Basic weight or substance number is 20
②	Determine square inches in basic size of sheet	②	17 x 22 = 374 sq. in.
③	Double basic weight and divide by square inches of basic size	③	20 x 2 = 40 40 ÷ 374 = .107 decimal weight factor Ans.

$17 \times 22 = 374 \overline{)40.000}$ = .107 decimal weight factor Ans.

- It is most helpful for the student to learn and understand how to use a paper dealer's catalog and price lists, since paper dealers determine the equivalent weight of paper in the required sizes other than the basic size.

- Printers may look for a decimal weight factor of the basic weight of the paper desired, Fig. 29-1. Once the factor is determined, it will be used (through multiplication) with the square inches of the sheet of paper to obtain the weight of 1,000 sheets.

> To find the equivalent weight of 1,000 sheets of same thickness and weight of paper, multiply dimensions of the desired sheet by the decimal standard number (DSN) for the weight (thickness) of paper desired.

DECIMAL WEIGHT FACTORS FOR BOND, BOOK, COVER PAPERS

BOND PAPERS	17 x 22 Basic (374 sq. in.) 1000 Sheets								
Basic Weight	16	20	24	28	32	36	40	44	49	51
DSN (1 M) (Decimal Standard Number)	.086	.107	.128	.150	.171	.193	.214	.236	.262	.272

BOOK PAPERS	25 x 38 Basic (950 sq. in.) 1000 Sheets								
Basic Weight	30	35	40	45	50	60	70	80	100	120
DSN (1 M)	.063	.074	.084	.095	.105	.126	.147	.168	.210	.252

COVER PAPERS	20 x 26 Basic (520 sq. in.) 1000 Sheets								
Basic Weight	25	35	40	50	65	80	90	100	130	160
DSN (1 M)	.096	.134	.154	.192	.250	.308	.346	.384	.500	.616

Fig. 29-1

In 1945, the National Paper Trade Association adopted the American Decimal Standard of Weights for Paper Stock (U.S. Bureau of Standards 1942). By adopting this system, the paper dealers factoring tables became standard throughout the industry.

Section 7 — Paper Stock

FORMULA: dimensions of sheet to be used × decimal weight factor for basic weight of paper = weight of 1,000 sheets of stock to be used on job.

Example: Find the equivalent weight of 1,000 sheets of 17″ x 28″ – 20 Bond paper.

① Find the decimal weight factor for the thickness and kind of paper stock desired (Fig. 29-1) ② Multiply dimensions of desired sheet by decimal weight factor for 1000 sheets	① Fig. 29-1 shows the weight factor for substance 20 Bond paper to be .107 ② 17 x 28 = 476 sq. in. 476 x .107 = 50.932 or 51 lbs. for 1000 sheets Ans.
17 x 28 x .107 = 50.932 or 51 lbs. for 1000 sheets Ans.	

- Apply the principles of equivalent weights of stock to the printing and graphic communications industry by solving the review problems.

REVIEW PROBLEMS

1. Determine the weight of a ream of paper 32″ x 44″ if the basic ream is 25″ x 38″ – 60. _____

2. Determine the weight of a ream of paper 35″ by 45″ if the basic ream is 25″ x 38″ and weighs 70 pounds. _____

3. Determine the weight of a ream of paper 22½ x 28½ if the basic ream is 20″ x 26″ – 65. _____

4. Determine the weight of a ream of paper 28″ x 34″ if the basic ream weighs 28 pounds and is 17″ x 22″. _____

5. Determine the weight of a ream of paper 17″ x 28″ if the basic ream weighs 16 pounds and is 17″ x 22″. _____

6. Determine the weight of a ream of paper measuring 19″ x 24″ if the basic ream weighs 20 pounds and is 17″ x 22″. _____

7. If a ream of paper 25″ x 38″ weighs 80 pounds, what will a ream 35″ x 45″ of the same grade weigh? _____

8. Determine the weight of a ream of paper 24″ x 36″ if the basic ream is 25″ x 38″ and weighs 60 pounds. _____

9. If the weight of a basic ream measuring 25″ x 38″ is 70 pounds, what is the weight of a ream 42″ x 58″? _____

10. If a ream of paper 17″ x 22″ weighs 20 pounds, find the weight of a ream of the same kind of paper, size 22″ x 34″. _____

11. What is the decimal weight factor for Bond paper substance 20? _____

Unit 29 — Equivalent Weights of Stock

12. Determine the weight of 3,000 sheets of Bond paper, size 28" x 34" — substance 20. _____

13. Determine the weight of 2,000 sheets of Book paper, size 24" x 30" — substance 80. _____

14. Determine the weight of 1,000 sheets of Book paper, size 35" x 45" — substance 70. _____

15. Determine the weight of a ream of paper 35" x 46" if the basic ream of paper measures 20" x 26" and weighs 80 pounds. _____

16. Determine the weight of 175 sheets of Bond paper 17" x 22" - 16. _____

UNIT 30 DETERMINING AND CUTTING PAPER STOCK

BASIC PRINCIPLES OF DETERMINING PAPER STOCK

- Prior to the printing of a job, the amount of paper stock needed must be determined and ordered. Often the printer must determine how many pieces of the estimated size of paper can be cut from a full-size sheet. The amount of stock needed for the particular job can then be estimated.

FORMULA: cuts in sheet   = $\dfrac{\text{dimensions of stock sheet}}{\text{dimensions of piece to be cut}}$

Example: How many pieces 6″ x 9″ can be cut from a sheet of 25″ x 38″ Book paper?

① Set up formula	① 25 ÷ 6 = ? 38 ÷ 9 = ?
② Divide vertically	② 25 ÷ 6 = 4 with 1 left over 38 ÷ 9 = 4 with 2 left over
③ Multiply quotients to get pieces	③ 4 x 4 = 16 pieces
④ Set up rough diagram and determine waste	④ 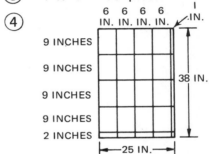 Waste is 1 piece 1″ x 38″ and 1 piece 2″ x 24″
⑤ Set up formula, but reverse the position of piece dimensions	⑤ 25 ÷ 9 = 38 ÷ 6 =
⑥ Repeat Division	⑥ 25 ÷ 9 = 2 with 7 left over 38 ÷ 6 = 6 with 2 left over
⑦ Multiply quotients to get pieces cut from stock	⑦ 2 x 6 = 12 pieces
⑧ Set up rough diagram and determine waste	⑧ 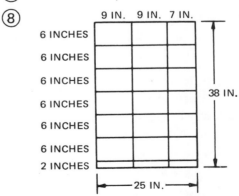 Waste is 1 piece 2″ x 18″ and 1 piece 7″ x 38″.

⑨ Determine if waste pieces can be used	⑨ 7" x 38" pieces can be cut 7 ÷ 6 = 1 38 ÷ 9 = 4 Thus, 4 pieces + 12 pieces = 16 pieces

⑩ Determine which cutting plan is best	⑩ Step 3 = 16 pieces
	           2" x 24" waste
	           1" x 38" waste
	Step 6 = 12 pieces
	           7" x 38" waste
	           2" x 18"
	The waste yields 4 pieces to give a total of 16 pieces
	The cutting plan in Step 3 is best because of strait cutting

⑪ Make cutting diagram (refer to Steps 1, 2, and 3)	⑪ 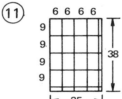 25 ÷ 6 = 4 with 1 left over 38 ÷ 9 = 4 with 2 left over
⑫ Number of pieces cut from a 25" x 38" sheet	⑫ Total of 16 pieces Ans.

BASIC PRINCIPLES OF CUTTING PAPER STOCK

- Sometimes, a sheet of stock can be turned in such a way that the waste cut away can be used to obtain more usable pieces from the sheet. However, when the printing must be done with the grain to assure rigidity (for book pages, covers, placards, posters, menus, programs, for offset and so forth), *strait cutting* (cutting with the grain) must be observed in determining all cuts.

- Paper folds more easily with the grain than across the grain. Books fold better and stay open when the grain runs up and down the page. Cardboard posters stand without sagging in store windows when the grain runs up and down rather than across the poster. Offset work requires that the grain run in the direction of the feeding of the paper stock. These factors should be considered when determining the cutting pattern for a sheet of stock.

- Apply the principles of determining and cutting paper stock to the printing and graphic communications industry by solving the review problems.

Section 7 — Paper Stock

REVIEW PROBLEMS

1. How many pieces 5″ x 3″ can be cut from one sheet of stock 22″ x 34″? _____

2. How many pieces 2½″ x 4½″ can be cut from one sheet of stock 28″ x 35″? _____

3. How many pieces 5″ x 7″ can be cut from one sheet of stock 25½″ x 30½″? _____

4. How many pieces 12½″ x 9½″ can be cut from one sheet of stock 38″ x 50″? _____

5. How many pieces 6″ x 9″ can be cut from one sheet of stock 20″ x 40″? _____

6. How many pieces 4″ x 6″ can be cut from one sheet of stock 24″ x 36″? _____

7. How many pieces 9″ x 11″ can be cut from one sheet of stock 36″ x 48″? _____

8. How many pieces 3½″ x 5½″ can be cut from one sheet 18″ x 23″? _____

9. How many pieces 6″ x 9″ can be cut from one sheet 25″ x 38″? _____

10. How many pieces 2½″ x 4¾″ can be cut from one sheet of 26″ x 40″ stock? _____

UNIT 31 DETERMINING THE NUMBER OF SHEETS REQUIRED

BASIC PRINCIPLES OF DETERMINING THE NUMBER OF SHEETS

> To find the number of stock sheets needed for a job, divide the number of cuts obtained from one sheet of stock into the number of press sheets that are needed to print the job.

FORMULA: number of stock sheets needed = $\dfrac{\text{number of copies wanted}}{\text{cuts obtained from one sheet of stock}}$

Example: How many sheets of stock should be requisitioned for an order calling for 8,500 copies of a job, if 5 pieces can be cut from one sheet of stock?

①	Number of copies wanted for job	①	8,500 copies wanted
②	Divide by the number of cuts obtained from 1 stock sheet	②	8,500 ÷ 5 = 1,700 stock sheets needed Ans.

$$\frac{8,500}{5} = 1,700 \text{ stock sheets Ans.}$$

> To find the number of stock sheets needed for a job consisting of many pages and signatures, divide the pages in one signature into the pages of the book; multiply the quotient by the number of copies wanted; and divide by the number of press sheets in one sheet of stock.

FORMULA: press sheets needed for one book = $\dfrac{\text{pages in the book}}{\text{pages in one signature}}$

press sheets   x   copies   =   press sheets   ÷   press sheets in
in one book   wanted   needed   one sheet of stock

=   stock sheets
    needed

Example: How many sheets of stock 25" x 38" are needed for a job of 4,000 copies of a book 6" x 9" containing 96 pages, printed in 16s sheetwise (16 pages in each form).

①	Pages in book	①	96 pages, size 6" x 9"
②	Pages in form	②	16 pages each form printed sheetwise
③	Pages in each signature	③	32-page signatures
④	Divide pages in signature into pages in book	④	96 ÷ 32 = 3 press sheets
⑤	Multiply by the number of copies wanted	⑤	3 x 4,000 copies = 12,000
⑥	Divide by number of press sheets in one sheet of stock	⑥	12,000 ÷ 1 = 12,000 stock sheets needed for job Ans.

$$\frac{96}{32} = \frac{3 \text{ press sheets} \times 4,000 \text{ copies}}{1 \text{ press sheet in one stock sheet}} = 12,000 \text{ stock sheets needed for job Ans.}$$

Section 7 — Paper Stock

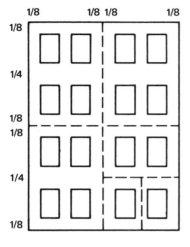

Fig. 31-1

Fig. 31-2

- Whenever paper stock is laid out for a bound book from which at least eight pages are obtained, an allowance of 1/8 of an inch must be added to the fore edge, head and tail of each page, as shown in Fig. 31-1. This allowance does not have to be added to the binding edge. This allowance is in addition to the margin already provided in the layout. After the stock is printed and folded, this allowance is trimmed from the three folded edges to insure pages of equal size in the finished book, as shown in Fig. 31-2.

Example:

Calculate the size of the press sheet needed for a 32-page signature if the basic page size is 6" x 9". If the basic page size is 6" x 9", add 1/8 inch to the 6-inch side (for the fore edge) and 1/4 inch to the 9-inch side (for the head and tail).

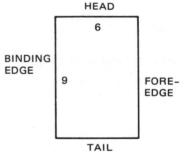

Thus, the basic page size is

6 1/8" x 9 1/4" for a 2-page signature (since both sides of the paper can be printed).

$$\frac{\times 2}{12\tfrac{1}{4}" \times 9\tfrac{1}{4}"}$$ for 4-page signature

$$\frac{\times 2}{12\tfrac{1}{4}" \times 18\tfrac{1}{2}"}$$ for 8-page signature

$$\frac{\times 2}{24\tfrac{1}{2}" \times 18\tfrac{1}{2}"}$$ for 16-page signature

$$\frac{\times 2}{24\tfrac{1}{2}" \times 37"}$$ for 32-page signature

Size of press sheet for a 32-page signature is 24½" x 37". Ans.

Unit 31 — Determining the Number of Sheets Required

- To eliminate needless calculations for the required paper stock, most printers use the table shown in Fig. 31-3. The numerals across the top of the table represent the number of press sheets (copies) desired. The numerals in the first column on the left represent the number of cuts (pieces) obtained from one sheet of stock. The numerals in the body of the table represent the number of sheets of stock required for the job.

    Example: The intersection of side number 8 with top number 500 indicates that 63 sheets of stock are needed for 500 copies (eight pieces out of one sheet).

- Apply the principles of determining the number of sheets required to the printing and graphic communications industry by solving the review problems.

CUTS	NUMBER OF PRESS SHEETS OR COPIES DESIRED											
	10	25	50	100	200	500	1000	1500	2000	2500	5000	10000
	NUMBER OF FULL SHEETS OF STOCK REQUIRED FOR A JOB											
1	10	25	50	100	200	500	1000	1500	2000	2500	5000	10000
2	5	13	25	50	100	250	500	750	1000	1250	2500	5000
3	4	9	17	34	67	167	334	500	667	834	1667	3334
4	3	7	13	25	50	125	250	375	500	625	1250	2500
5	2	5	10	20	40	100	200	300	400	500	1000	2000
6	2	5	9	17	34	84	167	250	334	417	834	1667
7	2	4	8	15	29	72	143	215	286	358	715	1429
8	2	4	7	13	25	63	125	188	250	313	625	1250
9	2	3	6	12	23	56	112	167	223	278	556	1112
10	1	3	5	10	20	50	100	150	200	250	500	1000
11	1	3	5	10	19	46	91	137	182	228	455	910
12	1	3	5	9	17	42	84	126	168	209	417	834
13	1	2	4	8	16	39	77	116	154	193	385	762
14	1	2	4	8	15	36	72	108	144	179	358	715
15	1	2	4	7	14	34	67	100	134	167	334	667
16	1	2	4	7	13	32	63	94	125	157	313	625
17	1	2	3	6	12	30	59	89	118	148	295	589
18	1	2	3	6	12	28	56	84	112	139	279	556
19	1	2	3	6	11	27	53	79	106	132	264	527
20	1	2	3	5	10	25	50	75	100	125	250	500
21	1	2	3	5	10	24	48	72	96	120	239	477
22	1	2	3	5	10	23	46	69	91	114	228	455
23	1	2	3	5	9	22	44	66	87	109	218	435
24	1	2	3	5	9	21	42	63	84	105	209	417
25	1	1	2	4	8	20	40	60	80	100	200	400
26	1	1	2	4	8	20	39	58	77	97	193	385
27	1	1	2	4	8	19	38	56	75	93	186	371
28	1	1	2	4	8	18	36	54	72	90	179	358
29	1	1	2	4	7	18	35	52	69	87	173	345
30	1	1	2	4	7	17	34	51	67	84	167	334
31	1	1	2	4	7	17	33	49	65	81	162	323
32	1	1	2	4	7	16	32	47	63	79	157	313
33	1	1	2	4	7	16	31	46	61	76	152	304
34	1	1	2	3	6	15	30	45	59	74	148	295
35	1	1	2	3	6	15	29	43	58	72	143	286
36	1	1	2	3	6	14	28	42	56	70	139	278
37	1	1	2	3	6	14	28	41	55	68	136	271
38	1	1	2	3	6	14	27	40	53	66	132	264
39	1	1	2	3	6	13	26	39	52	65	130	257
40	1	1	2	3	5	13	25	38	50	63	125	250

Fig. 31-3

Section 7 — Paper Stock

REVIEW PROBLEMS

1. How many pieces 5" x 6" can be cut from one sheet of stock measuring 17" x 28"? _____

2. How many pieces 5" x 8" can be cut from one sheet of stock measuring 25" x 38"? _____

3. How many sheets of 22" x 34" – 16 Bond are needed for a job of 1,200 letterheads 8½" by 11"? Show the cutting pattern for this problem. _____

4. A job order calls for 10,000 pieces 4¼" x 7" – 20. How many sheets of 17" x 28" Bond are needed? Show the cutting pattern for this problem. _____

5. How many sheets of stock are required to fill an order for 5,500 slips 3" x 4" to be cut from 19" x 24" – 20 Bond? _____

6. How many pieces of 2/3 letterhead size pages (7¼" x 8½") can be cut from one ream of 22" x 34" Bond? _____

7. a. How many sheets of 20½" x 24¾" – 91-pound Index Bristol are needed for a job of 12,000 pieces 5" x 8"? _____

   b. If a ream of this paper weighs 91 pounds, how much does the stock needed for the job weigh? _____

8. Find the amount of stock needed for 10,000 32-page booklets, trimmed size 6" x 9", cut from S.C. Book 38" x 50" — 80. _____

9. How much stock is needed for 15,000 15-page booklets, trimmed size 7 1/8" x 6 3/4", cut from stock 23" x 35" – 20? _____

10. Find the amount of stock necessary to print 15,000 32-page catalogs with the following specifications: trimmed size 6" x 6 7/8", cut from coated stock 38" x 50" – 70. _____

11. a. Find the number of 6" x 9" sheets that can be cut from 1,750 sheets of 25" x 38" – 80 stock. _____

    b. If one ream of this paper weighs 80 pounds, determine the weight of the order. _____

12. How many cards, size 3" x 5", can be cut from a lot of 2,000 scrap pieces of Index Bristol measuring 8" x 16"? _____

Unit 31 — *Determining the Number of Sheets Required*

13. An order is received to print 2,500 books consisting of 112 pages each on M.F. Book stock measuring 28" x 40". If the trimmed size is 6¾" x 9¾" and it is run in 16-page forms, how many sheets are needed?

14. How many reams of 17" x 22" bond paper are necessary to obtain 15,000 pieces measuring 5" x 7"?

15. A job is to be printed on a sheet 3½" x 3¼" and is to be cut from stock 28" x 44". How many pieces can be cut from one sheet of the 28" x 44" stock?

16. Billheads measuring 8½" x 5½" are to be printed on a Bond paper. How many reams of 17" x 22" stock are required to run 36,000 billheads?

17. Ten thousand billheads 8½" x 4¾" are ordered. How many sheets of 17" x 22" Bond are required to fill the order?

18. What is the cost of 2,500 covers 9" x 12" cut from cover stock 20" x 26" which costs $12.75 per 100 sheets?

19. a. A sheet of untrimmed paper measures 4½" x 6". Determine the size of the paper with the paper trim allowance.

    b. Find the size of the paper needed for a 32-page signature.

20. a. A sheet of untrimmed paper measures 5¼" x 7¼". Determine the size of the paper with the trim allowance.

    b. Find the size of the paper needed for an 8-page signature.

21. a. A sheet of untrimmed paper measures 6½" x 8¾". Determine the size of the paper with the paper trim allowance.

    b. Find the size of the paper needed for a 16-page signature.

UNIT 32 FINDING THE MOST ECONOMICAL CUT OF STOCK

BASIC PRINCIPLES OF ECONOMICAL CUTTING

- The opportunity of using several sizes of paper stock is always available to the printer. It is the printer's responsibility to determine the proper size to give the most desired units from a full size with the least amount of waste.

  When a printer has a choice of paper stock, the percentage of stock actually used is calculated. A comparison between paper sizes will determine the most economical choice.

> To determine which one of several sizes of stock is the most economical (most cuts, least waste, least cost), select two or more sheet sizes, and determine cuts from the sheet. When it is known how many cuts may be obtained from each sheet selected, multiply the dimensions of the cut piece by the number of pieces cut from the sheet, and divide by the dimensions of the sheet of stock from which they were cut.

FORMULA: $\dfrac{\text{pieces cut} \times \text{dimensions of cut pieces}}{\text{dimensions of stock sheet}}$ = area (in hundredths or percent) of sheet used

Example: Which stock sheet will cut 5" x 7" pieces most economically? Try three sizes of Bond paper: 17" x 22", 17" x 28", and 19" x 24."

① Figure cuts in each sheet A. from 17" x 22" stock B. from 19" x 24" stock C. from 17" x 28" stock	① Number of cuts 5" x 7" A. 10 pieces B. 11 pieces C. 12 pieces	
② Multiply cuts by dimension of cut pieces	② A. 10 x 5" x 7" = 350 B. 11 x 5" x 7" = 385 C. 12 x 5" x 7" = 420	
③ Divide by dimensions of stock sheet	③ A. 17" x 22" = 374 sq. in. 350 ÷ 374 = .935 or 94% B. 19" x 24" = 456 sq. in. 385 ÷ 456 = .844 or 84% C. 17" x 28" = 476 sq. in. 420 ÷ 476 = .88 or 88%	
④ Determine which is most economical	④ The first or the 17" x 22" stock is best Ans.	

A. $\dfrac{10 \times 5 \times 7}{17 \times 22} = \dfrac{\cancel{10}^{5} \times 35}{17 \times \cancel{22}_{11}} = \dfrac{175}{187} = .935$ or 94% Ans.

- Apply the principles of economical cutting to the printing and graphic communications industry by solving the review problems.

Unit 32 — Finding Most Economical Cut of Stock

REVIEW PROBLEMS

1. Which stock sheet will cut 5" x 6" pieces most economically: 17" x 22", 17" x 28", or 22" x 34"? _____

2. Which stock sheet will cut 5" x 7" pieces most economically: 18" x 46", 21" x 32", or 22" x 34"? _____

3. Which stock sheet will cut 4" x 6" pieces most economically: 18" x 23", 22" x 34", or 25" x 38"? _____

4. Which stock sheet will cut 8" x 10" pieces most economically: 25" x 38", 32" x 44", or 38" x 50"? _____

5. Which stock sheet will cut 4" x 7" pieces most economically: 28" x 42", 35" x 45", or 38" x 52"? _____

6. Which stock sheet will cut 8½" x 11" pieces most economically: 17" x 22", 17" x 28", or 21" x 32"? _____

7. Which stock sheet will cut 6¼" x 9½" pieces most economically: 17" x 22", 17" x 28", or 19" x 24"? _____

8. Which stock sheet will cut 5" x 6" pieces most economically: 25" x 38", 28" x 42", or 32" x 44"? _____

9. Which stock sheet will cut 7" x 9" pieces most economically: 25½" x 30½", 20½" x 24¾", or 22½" x 28½"? _____

10. Which stock sheet will cut 3" x 4" pieces most economically: 20" x 26", 23" x 35", or 26" x 40"? _____

UNIT 33 ALLOWANCE FOR PAPER SPOILAGE

BASIC PRINCIPLES OF DETERMINING THE ALLOWANCE FOR SPOILAGE

- Whenever printers estimate the amount of paper needed for a job, they must allow spoilage for cutting, make ready, registering, printing, binding, finishing and other operations. The printer usually adds additional sheets to an order to guarantee a full count on delivery to the customer. The Spoilage Percentage Table, shown in Fig. 33-1, is used by printers to assist in determining how much extra paper stock should be added to guarantee the full count.

PERCENTAGE OF PAPER ALLOWED FOR SPOILAGE

Number of finished copies desired	1-100	101-250	251-500	501-1000	1001-5000	5001-10000	10 M - 25 M	over 25 M
(Percentage of Spoilage Based on Press Sheets or Signatures)								
First time thru press	10	8	6	5	4	3	2½	2
*Each add'l impression	7½	6	4½	3¾	3	2¼	2	1½
Each bindery operation	5	4	3	2½	2	1½	1¼	1
Allow for 1st signature — sheetwise (w and b)	15	12	9	7½	6	4½	3¾	3
Work-and-turn (w and t)	14	11	8	6½	5½	4	3¼	2½
Allow for each add'l signature	5	4	3	2½	2	1½	1¼	1

*Included in additional impressions are considerations of separate plates, separate colors, finishing operations, the press and all other operations which require the sheets to be run through the press an additional time.

Fig. 33-1

To determine the number of extra sheets of stock that must be added to ensure a full count of copies when the job is delivered, find the percentage of spoilage and add these sheets to the stock sheets already determined for the job.

FORMULA: number of stock sheets needed   x   percentage of spoilage   =   extra sheets to be added for spoilage

Example: How many stock sheets should be cut for a job calling for 500 copies one time through the press (cut two from one sheet of stock) with allowance for spoilage?

Unit 33 Allowance for Paper Spoilage

①	Refer to the Cutting Table (Fig. 31-3) for cuts and stock sheets	①	2 out for 500 copies equals 250 stock sheets
②	Refer to Spoilage Table (Fig. 33-1) for percent of spoilage	②	6% one time through press
③	Multiply stock sheets by percentage of spoilage	③	250 x .06 = 15 extra sheets
④	Add extra sheets for spoilage	④	250 + 15 = 265 sheets needed Ans.

$$250 \times .06 = 15 + 250 = 265 \text{ sheets Ans.}$$

- Apply the principles of determining the allowance for spoilage to the printing and graphic communications industry by solving the review problems.

REVIEW PROBLEMS

1. What is the spoilage allowance on a job of 2,000 copies of an eight-page book printed work-and-turn? _____

2. How many additional sheets of stock 22½" x 28½" must be added to an order for 10,000 3" x 5" cards printed two colors, one side? _____

3. How many copies must be printed to cover spoilage for an order of 9,000 copies printed on two sides with one fold? _____

4. How many copies must be printed to cover spoilage for an order of 12,500 copies of a poster printed in three colors on one side? _____

5. How many copies must be printed to cover spoilage for an order of 1,000 copies of a bulletin printed on two sides with three colors on each side and three holes punched? _____

6. How many copies must be printed to cover spoilage for an order of 500 copies of a program card printed on one side in one color with one perforated side? _____

7. How many sheets of 25" x 38" M.F. will be needed to print 40,000 handbills 6" x 9" run in two colors and black on one side? (Include spoilage) _____

8. How many sheets of 20" x 26" cover stock will be needed to run 14,000 covers 9" x 12" in three colors and black, on two sides? (Include spoilage) _____

9. How many sheets of scrap material, 8" x 16", are needed to run 4,500 cards, 3" x 5", printed on both sides? (Include spoilage) _____

10. An order for 15,000 letterheads, 8½" x 11" are to be run on a sixteen-pound Bond. The size sheet to be purchased is 17" x 22". At 47½¢ per pound, what will the paper for the job cost, printed on one side? (Include spoilage) _____

Section 7 — Paper Stock

11. A job of 10,000 cards, 3½" x 5¼" is to be printed on both sides. If the cards are cut from 22½" x 28½" index, what does the stock cost at $9.50 per hundred sheets? (Include spoilage) _____

12. How many sheets of 20" x 26" stock are needed to print 42,000 sheets, 6" x 9", in two colors and black on one side? (Include Spoilage) _____

UNIT 34 CHARGING FOR CUTTING AND HANDLING STOCK

BASIC PRINCIPLES OF CHARGING

- The amount of time needed to cut paper stock is determined by the number of cuts per sheet and the number of sheets to be cut. Cutting charges are applicable to both the cutting of the paper stock in preparing it for press and at the completion of the job, prior to shipping. A sample price schedule for cutting paper stock is shown in Fig. 34-1.

CUTTING CHARGE PER 1000 STOCK SHEETS

Number of pieces cut from 1 sheet	Charge per M	Number of pieces cut from 1 sheet	Charge per M
2	$ .30	21-30	$1.80
3-4	.60	31-40	2.20
5-9	1.00	41-50	2.60
10-20	1.40	over 50	2.90

Fig. 34-1

- The maximum number of sheets of paper stock that is placed under the knife of a cutting machine, at one time, is called a *lift*. Listed below is a sample of various lifts that are handled on the cutting machine.

  500-sheet package for book and bond papers

  250-sheet package for covers (lightweight) and some blottings

  100-sheet package for cardboard and some cover papers
    (double thick, heavyweight)

  125-sheet package for blottings

- The cost of handling paper stock at any time during the printing of a job is considered an expense, and an additional charge must be included in the final cost. As a rule of thumb, most printers add 10 percent to the cost of the material to cover expenses such as shipping, trucking, storage, and jogging. Many printers include this cost in different areas of the final charge.

To determine the charge for cutting and handling stock, refer to a table for the charge per thousand stock sheets for cutting stock. Multiply this value by number of stock sheets determined for job (including spoilage) and divide by 1000. Add charge for preparation of cutter.

FORMULA: charge for cutting and handling stock = $\dfrac{\text{number of stock sheets} \times \text{cutting charge per 1000 sheets}}{1000}$ + preparation charge*

*Assume a charge of $.60 for the preparation of the cutter.

Example: What is the charge for cutting stock for a job order of 2,400 single-page programs, 6" x 9", printed on bond paper? (Include spoilage)

Section 7 — Paper Stock

①	Determine number of cuts to a sheet	①	5 cuts to sheet
②	Determine number of stock sheets needed for job	②	500 stock sheets needed
③	Refer to the table for the cutting charge (Fig. 34-1)	③	Table indicates a charge of $1.00 per M sheets
④	Multiply stock sheets needed by charge per thousand sheets	④	500 x $1.00 = $500.00
⑤	Divide by 1000	⑤	500 ÷ 1000 = $.500
⑥	Add preparation charge of $.60	⑥	$.50 + $.60 = $1.10 Ans.

$$\frac{500}{1000} \times \$1.00 = \frac{.500}{1000} \times 1 = \$.50 + \$.60 = \$1.10 \text{ Ans.}$$

- Apply the principles of charging to the printing and graphic communications industry by solving the review problems.

REVIEW PROBLEMS

1. What is the cutting charge for the stock for a job of 80,000 permit blanks, 4" x 6", cut from 17" x 22" stock? _____

2. What is the cutting charge for 1,000 sheets of stock cut 10 pieces to a sheet? _____

3. What is the cutting charge for 500 sheets of stock cut 30 pieces to a sheet? _____

4. What is the cutting charge for 7,000 sheets of stock cut 8 out of a sheet? _____

5. What is the cutting charge for 1,500 sheets of stock cut 20 out of a sheet? _____

6. What is the cutting charge for 750 sheets of stock cut 6 out of a sheet? _____

7. What is the cutting charge for 2,500 sheets of stock cut 30 out of a sheet? _____

UNIT 35 DETERMINING WEIGHT OF PAPER STOCK

BASIC PRINCIPLES OF PAPER STOCK WEIGHT

- All paper stock is sold by weight. The printer uses the Decimal Weight Factor Table to determine the weight of the cut pieces printed for the job.

> To find the weight of stock, multiply the total number of pieces cut from the stock used by the dimension of the cut piece, and by the decimal weight factor for 1000 sheets and divide by 1000.

FORMULA: Weight = $\dfrac{\text{total number of cut pieces} \times \text{dimensions of cut piece} \times \text{factor for 1000 sheets}}{1000}$ (Fig. 29-1, page 60)

Example: What is the weight of 800 pieces, 4" x 6", cut from Bond paper, substance 20?

①	Determine the total number of pieces cut from stock used	①	800 pieces
②	Determine dimensions of cut piece	②	4" x 6"
③	Determine weight factor for sub. 20 Bond paper for 1000 sheets (see Fig. 29-1, page 60)	③	Table in Fig. 29-1, page 60 shows: .107 for sub. 20
④	Multiply	④	800 x 4 x 6 x .107 = 2054.4
⑤	Divide by 1000	⑤	2054.4 ÷ 1000 = 2.05 + or 2 lbs. Ans.

$\dfrac{800 \times 4 \times 6 \times .107}{1000}$ = .800 x 24 x .107 = 2.05 + or 2 lbs. Ans.

Example: What is the weight of 265 stock sheets, 17" x 28", Sub 20?

①	Multiply number of stock sheets by dimensions of stock sheet used	①	17 x 28 = 476 sq. in. 476 x 265 = 126140
②	Multiply by weight factor for sub. 20 Bond paper for 1000 sheets (Fig. 29-1, page 60)	②	Weight factor for sub. 20 Bond paper is .107 126140 x .107 = 13496.98
③	Divide by 1000	③	13496.98 ÷ 1000 = 13.496 or 13½ lbs. Ans.

$\dfrac{265}{1000}$ x 476 x .107 = $\dfrac{.265}{1000}$ x 476 x .107 = 126.14 x .107 = 13.496 + or 13½ lbs. Ans.

Example: solve the previous problem without using the factoring table.

Section 7 — Paper Stock

①	Number of stock sheets used	①	265 stock sheets
②	Multiply by weight of 1 M sheets	②	Calculation by the square inch method (Unit 29) shows that 1000 sheets of 17" x 28" - 20 Bond paper weigh 51 lbs. 265 x 51 = 13,515
③	Divide by 1000	③	13515 ÷ 1000 = 13.5 + or 13½ lbs. Ans.

$$\frac{.265}{1000} \cancel{265} \times 51 = .265 \times 51 = 13½ \text{ lbs. Ans.}$$

- Apply the principles of paper stock weight to the printing and graphic communications industry by solving the review problems.

REVIEW PROBLEMS

1. Find the weight of 1,225 sheets of 19" x 24", substance 16 Bond paper. _____
2. Find the weight of 2,250 sheets of 24" x 38", substance 60 book paper. _____
3. Find the weight of 4,550 sheets of 38" x 50", substance 70 book paper. _____
4. Find the weight of 4,250 sheets of 28" x 34", substance 44 Bond paper. _____
5. Find the weight of 10,000 sheets of 25½" x 30½", substance 20 Bond paper. _____
6. Find the weight of 400 sheets of 20" x 26", substance 90 M cover stock. _____
7. Find the weight of 150 sheets of 25½" x 30½", substance 40 Bond paper. _____
8. Find the weight of 825 sheets of 25" x 38", substance 80 book paper. _____
9. Find the weight of 25,000 sheets of 20" x 28", substance 20 Bond paper. _____
10. Find the weight of 8,500 sheets of 28" x 44", substance 90 book paper. _____
11. Find the weight of 600 3" x 5" pieces cut from substance 20 Bond paper. _____
12. Find the weight of 150 4" x 6" pieces cut from substance 60 book paper. _____

UNIT 36 DETERMINING COST OF PAPER STOCK

BASIC PRINCIPLES OF PAPER STOCK COST

- All printers should be familiar with the use of the paper dealer's catalog and price lists. A typical list (5 column headings) is shown in Fig. 36-1.

2000 lbs. and over	500 lbs. to 1999 lbs.	125 lbs. to 499 lbs.	Full Pkg. Price	Brkn Pkg. Price
—	—	—or—	—	—
16 cartons 2000 lbs.	4 cartons 500 lbs.	1 carton 125 lbs.	Unit Pkg. Price	Brkn Pkg. Price

Fig. 36-1

- A penalty charge, ranging from 15-50 percent of the cost of the paper, is added whenever it becomes necessary to break a package. It is advisable that the printer use the right price and category in the paper catalog and price list, since the paper stock used for a job is the single most expensive item of the final printing cost.

- Unit prices in each column usually identify the cost of a single pound of paper or 100 pounds (cwt) of paper. The use of the decimal point (dividing by 100) is the procedure used by most printers to convert the price per cwt to the price per pound. Printers usually look for a price break by determining which column to use. Of those columns shown in Fig. 36-1, the categories most printers use fall between 4-16 cartons, provided there are no broken single cartons or broken multiples of four.

FORMULA:

$$\text{number of full pkgs. in 1 carton} = \frac{125 \text{ lbs. in carton}}{\text{lbs. in 1 full pkg. of stock}}$$

$$\text{number of full pkgs. used} = \frac{\text{number of sheets used}}{500 \text{ sheets in full pkg.}}$$

$$\text{weight of stock used} = \text{number of full pkgs. used} \times \text{weight of 1 full pkg.}$$

Example: Should the carton price be used for an order calling for 3,000 sheets of 17" x 22" Bond paper — substance 20?

① Refer to the Packing Schedule for cartons of Bond paper (Fig. 36-2, page 81)	① Packing schedule shows 3,000 sheets of Bond paper sub 20 are packed in 1 carton
Ⓐ① Divide by weight of full package into 125 lbs.	Ⓐ① 125 ÷ 20 = 125/20 = 6 + full packages
② Multiply weight of full package	② 6 x 20 = 120 lbs. (approx. 125 lbs. or 1 carton)
Ⓑ① Divide number of sheets used by number of sheets in full package	Ⓑ① 3000 ÷ 500 = 6 full pkgs.
② Multiply by weight of full package	② 6 x 20 = 120 lbs. (approx. 125 lbs. or 1 carton) Ans.
Use the *carton* price Ans.	

Section 7 — Paper Stock

- Printers use similar methods to determine when to use the 4-carton (case) price and the 16-carton (ton) price.

Anybrand BOND PAPER Sealed 500 sheets Substance 20					WHITE — Linen Finish		
Trade Sizes	1000 Sht. Wght.	Carton Contents (Sheets)	16 cartons per cwt.	4 cartons per cwt.	1 carton per lb.	500 sheets per lb.	Brkn. pkg. per lb.
17 x 22 22 x 34 24 x 38	40 80 98	3000 1500 1500	19.20	20.70	23.00	.33	.49

Fig. 36-2

- Paper dealers usually add a penalty charge of approximately 50 percent for breaking a package. Referring to Fig. 36-2, note that 50 percent of 33 cents (lb. rate for full pkg. of 500 sheets) is 16 cents; 33 cents plus 16 cents equals 49 cents, the rate listed in the last column of figure 36-2 for a broken package.

To find the cost of paper stock, multiply the number of sheets used (or wanted) by the weight of 1000 sheets and by the price per pound and divide by 1,000.

FORMULA: cost of paper = $\dfrac{\text{number of sheets used} \times \text{weight per 1000 sheets} \times \text{price per lb.}}{1{,}000}$

Example: What is the cost for 1,800 sheets of Bond paper, substance 20?

① Multiply number of sheets by weight of 1,000 sheets	① 1,800 x 40 = 72,000
② Divide by 1,000	② 72,000 ÷ 1,000 = 72
③ Multiply by price per lb. (see price list, Fig. 36-2)	③ 72 x $.33 = $23.76 Ans.

$\dfrac{1{,}800 \times 40}{1{,}000} \times \$.33 = \dfrac{\overset{72}{\cancel{1800}} \times 40 \times \$.33}{\underset{25}{\cancel{1000}}} = 72 \times \$.33 = \$23.76$ Ans.

NOTE: $\dfrac{1{,}800 \text{ sheets used}}{3{,}000 \text{ sheets in carton}}$ = less than 1 carton but more than 1 full package

$\dfrac{1{,}800 \text{ sheets used}}{500 \text{ sheets in carton}}$ = 3.6 full packages in order

Use the *full package* price for paper.

- When the price is for 100 or for 1,000, or any unit other than pounds, follow this rule:

To find the cost of card stock, envelopes, covers and so forth, priced by the 1,000, 5,000, or count, divide the number of sheets (units) wanted by the selling count unit and multiply by the charge per selling unit.

81

Unit 36 Determining Cost of Paper Stock

FORMULA: cost of stock = $\dfrac{\text{Sheets wanted} \times \text{cost per selling unit}}{\text{Selling count unit}}$

Example: How much will 75 sheets of 25½" x 30½" – 100 Index Bristol cost?

①	Determine number of sheets (units) wanted	①	75 sheets wanted
②	Divide by selling count unit (see price table, Fig. 36-3)	②	Bristol sold by 1000 75 ÷ 1000 = .075
③	Multiply by cost (see price table, Fig. 36-3)	③	75 sheets less than full pkg. (100 sheets). Therefore, use Broken Package Column $73.00 M sheets (see price list) .075 x $73.00 = $5.48 Ans.

$$\dfrac{75 \times \$73.00}{1000} = \dfrac{75 \times \$\overset{\$.073}{73.00}}{1\cancel{000}} = \$5.48 \text{ Ans.}$$

Anybrand INDEX BRISTOL 25½ x 30½ Grain Long Way				WHITE — Smooth Finish Sealed or marked 100 Sheets Priced per 1000 Sheets			
Basic Weight	Weight M Sheets	Carton Contents	16 Cartons	4 Cartons	1 Carton	100 Sheets	Less than 100 Sheets
90	180	800	25.00	27.00	30.00	45.00	60.00
110	220	600	31.00	33.00	37.00	65.00	73.00
140	280	500	39.00	42.00	42.00	70.00	93.00
170	340	400	50.00	54.00	60.00	90.00	120.00
220	440	300	62.00	67.00	75.00	120.00	145.00

Fig. 36-3

- Envelopes are packaged 500 to a box, are priced per 1000, and are usually sold in carton lots. A paper dealer's catalog will list envelopes as shown in Fig. 36-4.

Anybrand BOND ENVELOPES 500 to box — Not banded Regular finish — WHITE			Priced per 1000			
Size	Substance	M's per crtn.	10-19 crtns.	5-9 crtns.	1-4 crtns.	Less crtn.
6¼	20	5	2.99	3.33	3.71	4.56
6¾	20	5	3.12	3.48	3.87	4.77
6¾	24	5	3.52	3.93	4.87	5.38
10	20	2½	5.05	5.63	6.26	7.71
10	24	2½	5.67	6.33	7.04	8.68

Fig. 36-4

Section 7 — Paper Stock

To find the cost of envelopes, divide number of envelopes desired by **1000** and multiply by unit price per thousand in accordance with price list for amount ordered (and used).

FORMULA: cost of envelopes = $\dfrac{\text{number of envelopes} \times \text{unit cost per thousand}}{1000}$

Example: What is the cost of 2,500 envelopes, #10 White Bond, substance 20?

①	Determine number of envelopes ordered	①	2,500 envelopes
②	Determine if amount is equal to a carton or less than carton	②	1 carton = 2500 envelopes Use 1 carton price
③	Determine unit price	③	Catalog shows price to be $6.26 M
④	Divide envelopes wanted by 1000	④	2,500 ÷ 1000 = 2.5
⑤	Multiply by unit selling cost	⑤	2.5 x $6.26 = $15.65 cost of envelopes Ans.

$$\dfrac{2{,}500 \times \$6.26}{1000} = \$15.65 \text{ Ans.}$$

REVIEW PROBLEMS

1. If 6 reams of paper cost $45.00, what will 27 reams cost?
2. What is the cost of 375 sheets of Hammermill Bond 19″ x 24″ – 20, at 62¼ cents per pound?
3. Find the cost at 42 cents per pound of 8,500 sheets of Power Bond, size 19″ x 24″. The basic weight is 17″ x 22″ — 16.
4. If the basic weight is 25″ x 38″ — 70, find the cost at $.275 per pound of 30,000 sheets, 32″ x 44″.
5. What is the cost of 4,000 sheets, 3″ x 6¼″, cut from Adirondack Bond, 17″ x 22″ – 16, at 55 cents per pound?
6. How much does the stock cost for 20,000 slips of paper, 3″ x 5″, cut from Adirondack Bond 17″ x 22″ — 20, at 45 cents per pound?
7. What is the cost of 1,200 covers for a booklet, page size 5″ x 8″, cut from Hammermill Cover 20″ x 26″ – 65 at 57½ cents per pound?
8. A customer wants 10,000 cards, 3½″ x 5¼″. The cards are cut from Daily Sales Index, 20½″ x 24¾″. What is the total cost of stock needed at $6.05 per 100 sheets?
9. Find the cost of the stock needed for 15,000 sets of prints consisting of an original 5″ x 8″ printed on Adirondack Bond White 17″ x 22″ – 16 at $.44 per pound and a duplicate printed on Time Card Index 20½″ x 24¾″ – 91 at $9.85 per hundred sheets?

Unit 36 Determining Cost of Paper Stock

10. What is the cost of stock needed for 7,000 copies of a 32-page booklet if the trimmed page size is 6" x 9", cut from Eggshell Book paper, 25" x 38" – 70, at 49 cents per pound?

11. A job calls for 30,000 sets printed on white, canary and blue paper, whose size is 8½" x 11". Allowing 2 percent for spoilage, what is the cost of 17" x 22" – 16 Bond white at $.6495 per pound and canary and blue at $.6625 per pound?

12. What is the cost of 8,000 sheets, size 5½" x 8½", to be cut from 17" x 22" – 16 stock at 42 cents per pound?

13. What is the cost of the stock required for 4,000 copies of a 4-page booklet, page size 9" x 12", cut from Gainsboro Ivory, 25" x 38" – 80, at 45½ cents per pound?

14. Find the cost of the stock used to print 12,000 letterheads, size 8½" x 11". The stock used is 17" x 22" – 20 Bond and costs 42 cents per pound.

15. Find the cost of the stock needed to print 8,000 16-page programs, trimmed size 6" x 9", cut from M.F. Book stock 25" x 38" – 70. The stock costs 48½ cents per pound.

16. What is the cost of stock for 9,000 billheads, 5½" x 8½" cut from Management Bond 22" x 34" – 20 at $.4535 per pound?

17. A ream of Hammermill Bond 17" x 22" weighs 20 pounds. What is the cost of 1,500 letterheads, 8½" x 11", cut from this stock if the stock costs $12 per ream?

18. When ordering stock for a 12-page pamphlet, it is found that the paper required is M.F. Book 25" x 38" – 70, which is listed at 46 cents per pound. The size of each page, after trimming, is 6" x 9". The job calls for 7,500 pamphlets. What is the cost of the stock needed for this job?

19. Find the cost of the paper for each of the orders given in the chart.

	Basis	Weight	Stock	Order Sheets	Cost	Kind	Total Cost
a.	17 x 22	16	19 x 24	17500	42½¢	Bond	?
b.	24 x 36	35	28 x 42	25000	24¾¢	Newsprint	?
c.	25 x 38	80	32 x 44	42500	36½¢	Book	?
d.	25 x 38	70	28 x 42	27250	48¼¢	Book	?
e.	24 x 36	40	32 x 44	37565	25¼¢	Newsprint	?
f.	25 x 38	90	24 x 36	78725	47½¢	Book	?
g.	17 x 22	20	19 x 24	11510	47¼¢	Bond	?
h.	32 x 44	80	24 x 36	55200	36¾¢	Book	?
i.	25 x 38	70	24 x 38	37500	37¼¢	Book	?
j.	24 x 36	60	32 x 44	15825	24¼¢	Newsprint	?

Section 7 — Paper Stock

20. Find the cost of 9 reams of 16-pound paper, 17" x 28", at 45 cents per pound. _____

21. Find the cost of 125 sheets of 20-pound paper, size 17" x 22", at 42½ cents per pound. _____

22. What is the cost of 150 sheets of 24-pound paper, size 17" x 28", at $.49 per pound. _____

23. What is the cost of 250 sheets of 40-pound paper at 45 cents per pound? _____

24. What is the cost of 1,275 sheets of 60-pound paper, size 25" x 38", at 57½ cents per pound? _____

25. If the price per hundred pounds of bond paper is as follows, determine the cost of the quantities listed:

    a. 1,290 sheets, 18" x 22" – 16 _____

    b. 150 sheets, 19" x 24" – 16 _____

    c. 18 sheets, 21" x 32" – 16 _____

    d. 1,725 sheets, 23" x 36" – 16 _____

    e. 2,650 sheets, 34" x 44" – 16 _____

Sixteen Cartons	Four Cartons	One Carton	Full Pkg.	Broken Pkg.
$29.90	$30.25	$31.25	$34.75	$40.25

26. What do 7 reams and 125 sheets of bond, 17" x 22" – 16, cost at 39 cents per pound? _____

85

ACHIEVEMENT REVIEW A

1. A printer completes the following jobs in one day: 8,750 letterheads; 17,500 business cards; 12,500 circulars; and 19,725 labels. Determine the total number of pieces of work the printer completes during the day.

2. A typographer, using an automatic tape-driven, photo typesetter, set 1,875 lines on Monday, 2,400 lines on Tuesday, and 2,150 lines on Wednesday. Determine the total number of lines the typographer sets during the three days.

3. If a publisher purchases a carload of paper stock for $36,600 and sells it to another publisher for $45,350, determine the amount of profit the first publisher makes.

4. A litho-offset printing press can produce 4,500 impressions per hour. How many impressions can be produced if the press runs for 5½ hours?

5. A book to be printed has a total of 3,468 lines of 10-pt. type. How many pages will there be if there are 32 lines on each page?

6. The inventory of a stripping department consists of the following number of packages of stripping material: 4½ reams small size, 25¾ reams medium size, and 38¼ reams larger size. Determine the total number of reams of stripping material in the inventory.

7. Three artists work on designing a new package for their advertising agency. The first artist works 9¼ hours; the second, 9½ hours; and the third, 5¾ hours. Determine the total number of hours the three artists work on this job.

8. The following amounts of presswash are taken from a 55-gallon drum: 8¼ gallons, 4¾ gallons, 2½ gallons, and 6¼ gallons. Determine the amount of presswash remaining in the drum, in gallons?

9. From the diagram below, determine the area of the space, in square inches if its size is 8¼ inches wide by 4 inches deep.

8 1/4 INCHES

4 INCHES

Achievement Review A

10. If 2½ ounces of etching solution is needed for each gallon of water to maintain a proper pH for the fountain solution of an offset press, how many ounces of etching solution will be needed if 6 gallons of fountain solution must be mixed?

11. A shipment of photographic chemicals is received. The weight of each chemical is shown in pounds.

    Litho developer......................... 47.5
    Replenisher for automatic processors 92.3
    Hypo (fixer) 61.6

    Determine the total weight of the shipment.

12. The working time of a pressman during one week is as follows:

    Monday............................. 6.7 hours
    Tuesday............................. 5.6 hours
    Wednesday .... 7.4 hours
    Thursday............................ 5.8 hours
    Friday.............................. 5.9 hours

    Determine the total working time of the pressman.

13. A collar for a form roller of an offset printing press has an outside diameter, O.D., of 1.7246" and a wall thickness, W.T., of .327". Determine the inside diameter, I.D., of the collar.

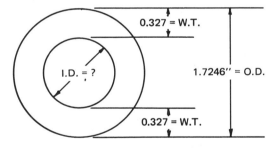

14. A printing job is billed for $2,345. The cost of the labor, materials and overhead are listed:

    Labor $1,140.50
    Materials 512.25
    Overhead...............................   172.00

    Determine the profit left after the labor, materials and overhead costs are paid.

15. An offset printing plant is owned by four persons. Mr. Jones owns 0.30 of the business, Mr. Adams owns 0.25, and Ms. Smith owns 0.35. The balance is owned by Ms. Thompson. What percentage of the business is owned by Ms. Thompson?

Achievement Review A

16. What is the thickness of 1,450 sheets of chipboard if each sheet is .125 inch thick?

17. What is the thickness of 8,750 sheets of index stock if each sheet is .009 inch thick?

18. A printing plant sells its scrap paper for 45.5 cents per pound. Determine the amount of money the plant receives if it sells 14,775 pounds of scrap paper.

19. If a one-pound can of ink costs $3.06, how many one-pound cans of ink can be purchased for $78.56?

20. If twenty-five gallons of offset roller wash weigh 165.7 pounds, determine the weight of one gallon of wash.

21. A bill for $980 for materials is submitted to a customer. Of this amount, 15% is the profit made by the dealer. Determine the profit of the dealer.

22. A paper merchant sells a quantity of paper at a profit of 13% and makes $450. What is the selling price of the paper?

23. The monthly bills owed by a customer are $1,927.72. If the bills are paid on or before the 10th of the next month, the customer is given a 3% discount on the bills. What is the total amount of the bills after the discount is taken?

24. The net price of the parts used to repair an offset press is $89.50, of which $52.00 is subject to a 33 1/3% discount from net price. What is the total price of the parts?

25. An offset duplicator whose price is $4,798.50 decreases in value 35% in one year. Determine the dealer's loss.

26. A stripping table costs $1,950. The dealer's profit is 25% on the cost of the table. Determine the dealer's profit.

27. A photo typesetting machine normally sells for $1,975. For a sale, the price of the machine is reduced 23%. Determine the sale price of the machine.

28. Determine the interest earned if $1,200 is left in a savings account for one year. The interest rate is 7% per year.

29. Determine the number of mm in fourteen cm.

30. Determine the number of cm in fifteen dm.

ACHIEVEMENT REVIEW B

1. The estimated value of a small offset duplicating plant is $19,250. Since the estimate, the value has decreased by $1700. Determine the present value of the plant.

2. If 1,000 sheets of book paper weigh 160 pounds, determine the weight of 22,500 sheets of the paper.

3. A typesetter can set an average of 350 lines of type per hour. Determine the number of lines of type the typesetter can set in 6 hours.

4. Two photo-typesetters work a total of 350 hours on a job. If they each work 7 hours a day, 5 days a week, how many weeks does the job take?

5. A publisher orders 52,500 pamphlets which are folded in a bindery by 4 workers. What is the average production of each worker?

6. The time sheet for printing a job, as shown below, indicates the amount of time needed for each operation.

    a. Determine the total amount of time needed for the job.

    b. What is the total time required if 6¾ hours for artwork are added to the time sheet?

TIME SHEET	
Camera	18¼ hours
Stripping and Platemaking	24¾ hours
Presswork	21½ hours
Cutting and Shipping	10¼ hours
Total Time	

7. The weekly classroom schedule for an apprentice is as follows: 6½ hours in a camera class, 2½ hours on stripping, and 4¼ hours in platemaking class. Determine the number of hours the apprentice spends in the three classes.

8. It takes 7½ hours to complete a job. A stripper works on the job during three different days as follows: ¾ hour, 1¼ hours, and 2¼ hours. How many hours are needed to complete the job?

9. Six lengths of photo-typesetting film, each 8½ inches long, are cut from a roll of film. How many inches of film are cut from the roll?

10. A folding machine can fold 4,000 sheets in an hour. How many sheets are folded if the machine runs 6¼ hours?

89

Achievement Review B

11. If a camera department uses 55 gallons of hypo (fixer) in 3½ days, what is the average amount of fixer used each day?

12. A certain company used 1,750 pounds of offset vellum paper. What is the total number of sheets used, if 1,000 sheets weigh 100 pounds?

13. Determine the total amount of the following bill:

Type and artwork	$ 47.75
Camera	28.50
Stripping	39.40
Platemaking	49.25
Presswork	425.75
Finishing	62.50
Tax	97.75

14. The printing paper stock in the storage room of an offset printing plant is valued at $9,972.50. During the week, $4,250.50 worth of paper is taken out of stock. Determine the value of the remaining stock.

15. The ink storage area of the pressroom had $1,775.50 worth of ink on its shelves. During one month, $625.50 worth of ink was used. Determine the value of the remaining ink.

16. What is the weight of 750 sheets of bond paper, if one sheet weighs .048 pound?

17. What is the cost of 72 pounds of offset black ink if the ink costs $2.25 per pound?

18. What does it cost to ship 79 cartons of printed materials if the cost to ship one carton is $1.25?

19. The cost of 830 inches of litho film is $12.72. Determine the cost of one inch of film.

20. A customer's bill is $3,975.00. The customer pays the printer one half of the bill in cash and pays the remainder of the bill in four equal payments. What is the amount of each of the equal payments?

21. A printer pays $947.50 for the stock, $224.50 for cold-type composition, and $26.00 for the miscellaneous labor on a job. The printer charges 15% profit on stock, 10% on composition, and 30% on the miscellaneous labor. Determine the total amount of the printer's bill for the job.

22. A quick copy center sells a printing job for $225.00. The loss on the job is 9%. What is the cost of the job to the copy center?

23. Two discounts are given on a bill of $2,450. The first discount is 30% and the second discount is 10%. Determine the final amount of the bill.

Achievement Review B

24. An agent receives a $175 commission for selling a $2,400 film developing sink. Determine the commission rate the agent receives. _____

25. A camera sells for $8,950. A commission of 15% on the sale is given to the salesperson. Determine the amount of commission the salesperson receives. _____

26. The federal income tax on a lithographer's payroll earnings is 24%. Determine the tax withheld on a paycheck of $225. _____

27. Rubber material for flexographic printing plates costs $3.95 per square foot. What is the cost of a piece 5 feet long and 42 inches wide? _____

28. The owner of a store which is 42 feet long and 33 feet wide, charges $2.75 per square foot per month rental. What is the rent per month? _____

29. The flat rate price for the labor on a printing job is $74. The job takes 5½ hours. What is the hourly cost for labor? _____

Contributions by Delmar Staff

Publications Director — Alan N. Knofla

Source Editor — Marjorie A. Bruce

Project Editor — Mitchell T. Baer

Director of Manufacturing/Production — Frederick Sharer

Production Specialists — Sharon Lynch, Jean LeMorta, Patti Manuli, Betty Michelfelder, Debbie Monty, Alice Schielke, Lee St. Onge

Illustrators — Tony Canabush, George Dowse, Michael Kokernak

The source editor would like to thank Elinor Gunnerson, Public Service Series Editor, for her careful review and proofing of the manuscript.

This edition of Practical Problems in Mathematics for Printers was prepared by Dr. James P. DeLuca, presently the chairman of the Department of Graphic Arts and Advertising Technology and the Lithographic-Offset Technology Program of the New York City Community College of the City University of New York.

Dr. DeLuca began his career in graphic communications as an apprentice platemaker. He gained full journeyman status as a combination letterpress/offset pressman. He later held positions as foreman, plant superintendent, and company manager of printing plants servicing advertising agencies. He is a member of the New York Printing Pressman's Union No. 51 and is active in the educational programs of the major unions servicing New York City. In 1960 Dr. DeLuca embarked on his academic career as a full-time instructor in graphic communications.

Dr. DeLuca received his high school vocational training at New York School of Printing and the Technical Trade School in Tennessee. By attending evening school, he obtained the A.A.S. degree in Graphic Arts and Advertising Technology. He received the B.S. degree in Industrial and Vocational Education (cum laude) from New York University, School of Education, and the M.S. degree (again cum laude) one year later from New York University. Dr. DeLuca received the Ed.D. degree in Community College Education from Nova University.

ANSWERS TO ODD-NUMBERED REVIEW PROBLEMS

SECTION 1 WHOLE NUMBERS
Unit 1 Addition of Whole Numbers

1. 43,925 pieces
3. 126 in.
5. 7,705 lbs.
7. 82 hr.
9. 5,045 papers
11. 32 items
13. a. 4 in.
    b. 6 in.
    c. 10 in.
15. 5,625 lines

Unit 2 Subtraction of Whole Numbers

1. 38 gallons
3. $108.00
5. a. 4 in.
   b. 7 in.
7. 28 hr.
9. 4,800 impressions

Unit 3 Multiplication of Whole Numbers

1. 21,000 impressions
3. 10,500 lb.
5. 35,000
7. 126,000 lb.
9. 360,000 lb.

Unit 4 Division of Whole Numbers

1. 25 reams
3. $.64
5. 72 pages
7. 30 hr.
9. 138 boxes

SECTION 2 FRACTIONS
Unit 5 Addition of Fractions

1. 26 3/4 hr.
3. 69 1/2 reams
5. 12 1/2 gallons
7. 5 1/16 in.
9. 25 1/4 hr.
11. 8 1/4 hr.
13. 39 1/4 gallons

Unit 6 Scale Reading

1. 48 divisions
3. 4 divisions
5. 14 divisions
7. 2 divisions
9. a. 3/4 in.
   b. 1 1/8 in.
9. c. 1 7/16 in.
   d. 1 13/16 in.
   e. 2 3/8 in.
   f. 2 15/16 in.
   g. 3 1/4 in.
   h. 3 11/16 in.
9. i. 4 1/8 in.
   j. 4 11/16 in.
   k. 5 9/16 in.
11. 2 1/2 in.
13. 1 7/16 in.

Unit 7 Subtraction of Fractions

1. 189 1/2 in.
3. 5/12 ream
5. 5 3/4 hr.
7. 6 3/4 lb.
9. 38 1/4 gallons

Unit 8 Multiplication of Fractions

1. $10.25
3. 297 1/2 lb.
5. 31,875 sheets
7. 1,239 words
9. 28 1/2 in.
11. 43 1/2 in.

Unit 9 Division of Fractions

1. 4,000 sheets/hr.
3. 17 ft/hr.
5. a. 5/6 in.
   b. 4 in.
7. a. 4+, therefore 4 boards
   b. 1 1/2 in.
9. 5 in.

SECTION 3 DECIMALS
Unit 10 Addition of Decimals

1. 16.0 hr.
3. 1.0663 in.
5. 10.55 in.
7. a. 36.625 in.
   b. 36.1 in.
   c. 37.245 in.
9. $6.75
11. $25.20

Answers to Odd-Numbered Review Problems

Unit 11 Subtraction of Decimals

1. 1.0706 in.
3. 9.45 hr.
5. $270.25

7. $597.25
9. a. 3.087 in.
   b. 1.65 in.

9. c. 3.861 in.
11. 1.0 hr.

Unit 12 Multiplication of Decimals

1. 156.25 in.
3. $3,989.37

5. $85.68
7. $623.00

9. $272.25
11. 45.94 or 46 cents

Unit 13 Division of Decimals

1. $11.39
3. 0.12 in.
5. 55.6
   a. 0.327 or 32.7%

5. b. 0.246 or 24.6%
   c. 0.426 or 42.6%
7. 22 cans

9. 6.984 lb.
11. $.0275
13. 277 lb.

Unit 14 Reading an Outside Micrometer

1. 6, 3, 0
3. 1, 3, 1.5
5. 4, 2, 0
7. 0, 0, 09

9. 0, 3, 13
11. 6, 0, 12
13. 9, 0, 12

15. 3, 2, 11
17. 6, 1, 14
19. 2, 2, 18

SECTION 4 PERCENT

Unit 15 Fractional Equivalents

1. a. 3/4
   b. 3/10
   c. 3/5
   d. 7/20
   e. 2/5
   f. 1/8

1. g. 2/3
   h. 1/6
   i. 1/4
   j. 3/8
3. 12 1/2%
5. 11.1%

7. 25%
9. 100 bottles
11. 10 bottles
13. a. 18.75%
    b. 81.25%

Unit 16 Simple Percent

1. 7.06%
3. $117.08
5. $1,644.50

7. 8%
9. a. $6,000.00
   b. $105.00

11. $281.70
13. $550.00
15. $940.43

Unit 17 Discounts

1. a. $1,529.70
   b. $1,483.81
3. $268.91

5. $453.60
7. a. $105.25
   b. $103.14

9. $1,371.85
11. $53.50

Unit 18 Profit and Loss, Commissions

1. a. $1,500.00
   b. 25%
3. $2.45

5. $655.50
7. $312.50

9. a. $28.35
   b. $1.42

Unit 19 Interest and Taxes

1. 3%
3. 3.25%
5. $4,674.71

7. $95.98
9. $15,060.87
11. $1,956.78

13. a. $48.00
    b. 8.7%

Answers to Odd-Numbered Review Problems

SECTION 5 MEASUREMENT

Unit 20 Linear Measure

1. 76.032 ft.
3. 205,920 in.
5. 3 1/6 ft.
7. 5,280 ft.
9. 6,160 yards
11. 18.94%
13. a. 72 in.
    b. 18 in.
    c. 38 1/4 in.
    d. 27 7/16 in.
15. a. 13.5 ft.
    b. 8.48 ft.
15. c. 4.19 ft.
    d. 5.38 ft.
17. 8 ft. 10 7/8 in.
19. 36 ft.
21. 16 cars

Unit 21 Metric System

1. 1,500 mm
3. 30,000 dm
5. 143.07 m
7. a. 18 m^2
7. b. 36 m^2
   c. 48 m^2

Unit 22 Metric Equivalent

1. 7.88 in.
3. 761.42 cm
5. Therefore, 12 1-m long pieces can be cut with a piece 0.8 m long remaining
7. 0.11 mm

Unit 23 Angular Measurement

1. 90°
3. 6 angles
5. 12 angles
7. 41° 18′
9. 34° 35′

Unit 24 Units of Area and Volume Measure

1. a. 25,000 sq. ft.
   b. 0.574 acres
3. $48.30
5. 41.66 or 41 presses
7. $1,500.00 per month
9. 6.800 lb.
11. 340 lb.
13. a. 7.481 gallons
    b. 6.232 gallons
15. 12 U.S. gallons

Unit 25 Time and Money Calculations

1. $28.91
3. 24 jobs
5. 14 days
7. 360 minutes
9. $5.25 per hr.

SECTION 6 GRAPHS

Unit 26 Practical Applications of Graphs and Charts

1. a.

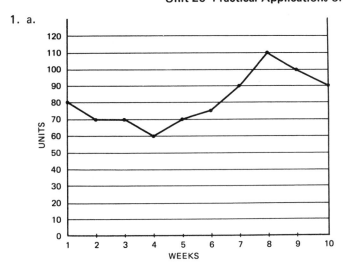

1. b. 81.5
   c. 4 weeks
   d. 6 weeks
3. a. $125.00
   b. $100.00
   c. $120.00
5.

Answers to Odd-Numbered Review Problems

Unit 26 Continued.

7.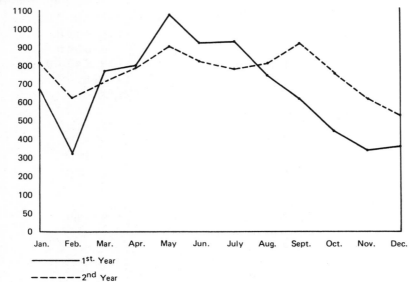

SECTION 7 PAPER STOCK

Unit 27 Packaging Paper Stock

1. 3 pkg.
3. a. 7 pkg.
   b. 14 pkg.
5. a. 7 pkg.
5. b. 2.33 ctn.
7. a. 35 pkg.
   b. 4.375 ctn.
7. c. yes, since 1 case = 4 cartons
9. a. 4 cartons
   b. 16 cartons

Unit 28 Basic Size, Thickness, and Weight of Stock

1. 17" x 22"
3. 20" x 26"
5. An underlined number in the dimensions of the paper indicates that the grain runs in the direction of this dimension
7. 140 lb.

Unit 29 Equivalent Weights of Stock

1. 89 lb. for 500 sheets
3. 80 lb. for 500 sheets
5. 20 lb. for 500 sheets
7. 133 lb. or 132 lb. for 500 sheets
9. 179 lb. for 500 sheets
11. .107
13. 242 lb. for 2,000 sheets
15. 248 lb. for 500 sheets

Unit 30 Determining and Cutting Paper Stock

1. 46 pieces
3. 20 pieces
5. 12 pieces
7. 16 pieces
9. 16 pieces

Unit 31 Determining the Number of Sheets Required

1. 14 pieces
3. 8 pieces per sheet
5. 145 sheets needed
7. a. 1,000 sheets needed
   b. 182 lb.
9. 7,500 sheets
11. a. 28,000 sheets
    b. 280 lb.
13. 8,750 sheets
15. 104 pieces
17. 1,250 sheets
19. a. 4 5/8" x 6 1/4"
    b. 18 1/2" x 25"
21. a. 6 5/8" x 9"
    b. 26 1/2" x 18" or 13 1/4" x 36"

Answers to Odd-Numbered Review Problems

Unit 32 Finding Most Economical Cut of Stock

1. Either the 17" x 22" or the 17" x 28" stock can be used.
3. Most economical cut is obtained by using 22" x 34" stock.
5. Most economical cut is obtained by using 28" x 42" stock.
7. Most economical cut is obtained by using 17" x 22" stock.
9. Most economical cut is obtained by using 25 1/2" x 30 1/2" stock.

Unit 33 Allowance for Paper Spoilage

1. 520 extra sheets for spoilage
3. 9,608 copies
5. 1,263 copies
7. 2,625 sheets
9. 602 sheets
11. $31.26

Unit 34 Charging for Cutting and Handling Stock

1. $.60
3. $.63
5. $.71
7. $.75

Unit 35 Determining Weight of Paper Stock

1. 48 lb.
3. 1,271 lb.
5. 832 lb.
7. 25 lb.
9. 1,498 lb.
11. 1 lb.

Unit 36 Determining Cost of Paper Stock

1. $204.75
3. $139.29
5. $4.16
7. $7.48
9. $149.53
11. $483.35
13. $145.60
15. $543.20
17. $9.00
19. a. $290.18
    b. $589.53
    c. $3,678.58
    d. $2,278.64
    e. $1,236.58
    f. $6,121.65
    g. $265.23
    h. $1,991.72
    i. $1,877.40
    j. $750.46
21. $2.13
23. $9.00

Achievement Review A

1. 58,475 pieces
3. $8,750.00 profit
5. 108.3 or 109 pages
7. 24 1/2 hr.
9. 33 sq. in.
11. 201.4 lb.
13. 1.0706 in.
15. 10%
17. 78.75 in.
19. 25 cans
21. $147.00
23. $1,869.89
25. $1,679.48
27. $1,520.75
29. 140 mm

Achievement Review B

1. $17,550.00
3. 2,100 lines
5. 13,125 pamphlets
7. 13 1/4 hr.
9. 51 in.
11. 15.714 gallons
13. $750.90
15. $1,150.00
17. $162.00
19. $.0153
21. $1,370.37
23. $1,543.50
25. $1,342.50
27. $69.13
29. $13.45

ACKNOWLEDGMENTS

Contributions by Delmar Staff

Publications Director — Alan N. Knofla

Source Editor — Marjorie A. Bruce

Project Editor — Mitchell T. Baer

Director of Manufacturing/Production — Frederick Sharer

Production Specialists — Sharon Lynch, Jean LeMorta, Patti Manuli, Betty Michelfelder, Debbie Monty, Alice Schielke, Lee St. Onge

Illustrators — Tony Canabush, George Dowse, Michael Kokernak

The source editor would like to thank Elinor Gunnerson, Public Service Series Editor, for her careful review and proofing of the manuscript.

This edition of Practical Problems in Mathematics for Printers was prepared by Dr. James P. DeLuca, presently the chairman of the Department of Graphic Arts and Advertising Technology and the Lithographic-Offset Technology Program of the New York City Community College of the City University of New York.

Dr. DeLuca began his career in graphic communications as an apprentice platemaker. He gained full journeyman status as a combination letterpress/offset pressman. He later held positions as foreman, plant superintendent, and company manager of printing plants servicing advertising agencies. He is a member of the New York Printing Pressman's Union No. 51 and is active in the educational programs of the major unions servicing New York City. In 1960 Dr. DeLuca embarked on his academic career as a full-time instructor in graphic communications.

Dr. DeLuca received his high school vocational training at New York School of Printing and the Technical Trade School in Tennessee. By attending evening school, he obtained the A.A.S. degree in Graphic Arts and Advertising Technology. He received the B.S. degree in Industrial and Vocational Education (cum laude) from New York University, School of Education, and the M.S. degree (again cum laude) one year later from New York University. Dr. DeLuca received the Ed.D. degree in Community College Education from Nova University.